Developing and Growing the Business – Freight

IMechE
Seminar Publication

I MECH E

Developing and Growing the Business – Freight

International Railtech Congress '98

24–26 November 1998
National Exhibition Centre,
Birmingham, UK

Organized by the Railway Division of the
Institution of Mechanical Engineers (IMechE)

In association with the:

Institution of Railway Signal Engineers
Institution of Civil Engineers
Railway Industry Association
Unione Internationale des Transport Publics
Institution of Electrical Engineers
UNIFE
Japan Society of Mechanical Engineers

IMechE Seminar Publication 1998–18

**Professional
Engineering
Publishing**

Published by Professional Engineering Publishing Limited for
The Institution of Mechanical Engineers, Bury St Edmunds and London, UK.

First Published 1998

ISSN 1357–9193
ISBN 1 86058 183 8

A CIP catalogue record for this book is available from the British Library.

Printed and bound in Great Britain by Bookcraft (Bath) Limited.

Related Titles of Interest

Title	Editor/Author	ISBN
New Trains for New Railways	IMechE Conference 1998–7	1 86058 146 3
Train Maintenance Tomorrow and Beyond	IMechE Conference 1997–1	1 86058 095 5
Better Journey Time – Better Business	IMechE Conference 1996–8	0 85298 997 0
Implementing Rail Projects	IMechE Conference 1995–1	0 85298 948 2
Fault Free Trains – A Reality	IMechE Seminar 1998–8	1 86058 133 1
Railway Traction and Braking (Railtech '96)	IMechE Seminar 1996–19	1 86058 018 1
Design, Reliability, and Maintenance for Railways (Railtech '96)	IMechE Seminar 1996–18	1 86058 017 3
Railway Rolling Stock (Railtech '96)	IMechE Seminar 1996–17	1 86058 016 5
Railway Engineering, System and Safety (Railtech '96)	IMechE Seminar 1996–16	1 86058 015 7

For the full range of titles published by Professional Engineering Publishing contact:

Sales Department
Professional Engineering Publishing Limited
Northgate Avenue
Bury St Edmunds
Suffolk
IP32 6BW
UK

Tel: 01284 724384
Fax: 01284 718692

Professional Engineering Publishing

Professional Engineering Publishing Limited is the new name for the publishing house of the Institution of Mechanical Engineers.

The reasons that we have changed our name after 25 years, from Mechanical Engineering Publications, are quite simple. As our readership and publishing aims grow, we need a name which reflects our plans for the future and our wider audience. Engineering is becoming a more complex and involved discipline; the boundaries between the various branches are becoming progressively blurred. There are increasing publishing opportunities for all our areas of interest including Journals, Newsletters, Electronic Products, Books and Magazines.

Professional Engineering Publishing is a leading international engineering publisher, the publishing house of the Institution of Mechanical Engineers (IMechE) and the exclusive European Agent for The American Society of Mechanical Engineers (ASME).

Professional Engineering Publishing will serve the international engineering community by expanding our publishing activities for the professional reader.

The history of mechanical engineering
A past that has built our future

An Engineering Archive

A Selection of Papers From the Proceedings of the Institution of Mechanical Engineers

Edited by Professor Desmond Winterbone

Archivist Keith Moore

An Engineering Archive celebrates the 150th anniversary of the foundation of the Institution of Mechanical Engineers and offers an excellent opportunity to look back at the contribution made by the Proceedings of the Institution to the whole area of mechanical engineering.

This volume contains a selection of papers, dating back to 1848, which have been chosen from the first 100 years of the Proceedings to exemplify both landmark accounts of engineering technology and science, and contributions from famous authors.

The papers reproduced are in facsimile, to show their original form. A number of the most memorable plates are included together with biographies of key engineering innovators.

300x210mm / 648 pages / 1997
Leather Bound:
ISBN 1 86058 052 1   £95.00
Quality Hardcover:
ISBN 1 86058 053 X   £59.00

Progress through Mechanical Engineering

By John Pullin

Published to mark the 150th anniversary of the founding of the world's first mechanical engineering institution, this book, with a foreword by HRH The Prince of Wales, charts the progress of the profession and the influences that have shaped engineering. At the same time it shows the key role that engineers and engineering have had, and will continue to have, on the world we live in, and the quality of life we lead, and illustrates the part that the Institution of Mechanical Engineers has played in those developments.

This volume will appeal to professional engineers, academics, those involved with the history and philosophy of science as well as readers with a more general interest.

1899163 28 X / 292x197mm / Hardcover / 288 pages / 1997
£28.00

Note: free delivery in the UK. Overseas customers please add 10% for delivery.

Credit card orders welcome

For further information:

Telephone Hotline - (24 Hour Service) - +44 (0) 1284 724384
Orders and enquiries to:
Sales Department, (eng), Professional Engineering Publishing,
Northgate Avenue, Bury St Edmunds, Suffolk, IP32 6BW, UK.
Fax: +44 (0) 1284 718692   E-mail: sales@imeche.org.uk

Professional Engineering Publishing

New Book Titles

Designing Cost-Effective Composites

This volume, based on the international conference held in London, UK, 15-16 September 1998, presents papers that are drawn from a range of experience and research into the uses, development, and analysis of composites.

It is an invaluable volume for design engineers, materials scientists, technologists, engineers in the fields of manufacturing, power and process industries, aerospace, automotive, and marine engineering. Researchers in academe, as well as those in government or independent laboratories, will also find this a useful source of information.

1 86058 148 X / 234x156mm / Hardcover / 291 pages / 1998 / £69.00

Forging and Related Technology (ICFT '98)

This IMechE conference volume addresses the fundamental and practical issues which are crucial to the successful production and optimum utilization of forged parts on a global basis. It is essential reading for all forward-looking manufacturers, designers, value analysers, material technologists, and users in the forging supply and related industries.

1 86058 144 7 / 234x156mm / Hardcover / 464 pages / 1998 / £124.00

Design Reuse
Engineering Design Conference '98

Edited by Dr S Sivaloganathan and Dr T M M Shahin

This volume containing appers presented at a conference organised by Brunel University in June 1998.

Drawing on the diverse experience of different sectors of industry and academic research, this prestigious collection of papers examines the ways in which the expertise in design can be structured and reused to carry the design process forward.

1 86058 132 3 / 234x156mm / Hardcover / 732 pages / 1998 / £189.00

Professional Engineering Publishing

For further information:
Telephone Hotline - (24 Hour Service) - +44 (0) 1284 724384
Orders and enquiries to:
Sales Department, Professional Engineering Publishing, Northgate Avenue, Bury St Edmunds, Suffolk, IP32 6BW, UK.
Fax: +44 (0) 1284 718692   E-mail: sales@imeche.org.uk

Contents

C552/045/98

Piggyback – meeting the customers' needs

T BERKELEY MA, MICE, FRSA, FCIT
The Piggyback Consortium, Oxford, UK

SYNOPSIS

This paper describes the development of piggyback in the UK, the equipment, the infrastructure enhancement required, and the potential market for new types of services. It demonstrates the links between the UK and continental systems and reports on the start of services in the UK using low height trailers. It describes the state of negotiations with Railtrack over the proposed gauge enhancement to carry 4 metre high trailers, and suggests that £100 million could be saved from Railtrack's current budget of £250 m by choosing an alternative diversionary route to avoid upgrading the West Coast Main Line fast tracks.

1. BACKGROUND

1.1 UK Rail Freight

After many years of decline, rail freight in the UK is now growing and last year achieved a 12% increase in tonne-miles carried by rail. Traditional markets such as aggregates, coal and steel remain strong, and new markets such as forest products, food and drink, waste disposal and construction materials are developing. The industry forecasts a tripling of rail freight in 10 years and the Rail Freight Group believes that, in 20 years, rail freight could have a 20% market share of freight traffic, compared with 7% today.

New markets and competitive pressures demand new solutions, and the greatest growth is forecast to be in unit loads rather than block trains, and in equipment and services most likely to be attractive to the customer. The new rail freight industry will be customer rather than supplier driven.

Within this context, the Piggyback Consortium has been developing over the last five years the idea that standard lorry semi-trailers could be carried on rail wagons in the UK. This not only relieves road congestion, but enables road hauliers to use rail for part or all of their journey with minimum new investment in equipment, and provides a more reliable trunk haul for industry in an era of growing congestion.

1.2 US experience

Piggyback services have operated successfully in the US and on the continent for many years. In the UK, services started this year, between Glasgow and London by EWS Railway for Parcelforce and between Manchester and Tilbury in October by Freightliner for Exel Logistics Tankfreight.

These services carry only low height van trailers or tanks because of the small loading gauge height on Britain's railways. Although there is a good niche market for such low height trailers, the majority of freight is carried in trailers that exploit the maximum available on road and cannot go by rail at the moment.

1.3 Gauge enhancement in the UK

The Piggyback Consortium has therefore been pressing Railtrack to raise the rail gauge height to allow 4 metre high trailers to be carried piggyback in the UK (Piggyback or PB Gauge, Fig 2). This height is compatible with the GB+ rail gauge common in much of continental Europe and many continental countries have a 4-metre height limit on road vehicles.

The Consortium commissioned a consultants study which indicated that, if the rail gauge height was raised between Glasgow and the Channel Tunnel, some 400,000 cross-Channel lorry journeys could transfer to rail, with domestic traffic adding further to this figure. The capital cost of raising the gauge to PB gauge was estimated in 1994 at under £100 million. Further studies commissioned by Railtrack from Deloitte Touche generally confirmed the overall demand for high gauge traffic.

Railtrack has chosen the West Coast Main Line (Fig 1) for this gauge upgrade and is investigating how the work can be integrated with the WCML passenger upgrade. They estimate higher costs, to include disruption costs to other operators, and confirm that, if piggyback gauge upgrade is to go ahead, a decision must be made by the end of 1998 in order to integrate the work with other possessions on the line necessary for the West Coast Main line upgrade between now and 2005.

2. WHAT IS PIGGYBACK?

Piggyback is the carriage by rail of lorry semi-trailers. The semi-trailers are generally lifted on and off the rail wagons, and secured by their 'fifth wheel' coupling to the wagon. The trailers are virtually standard with full width wheels, but must incorporate lifting points and be plated to ensure compliance with railway safety standards.

Trailers are delivered by road to terminals for loading onto rail for the trunk haul of the journey. The process is reversed at the destination terminal.

Piggyback is one of a variety of freight systems which have been developed to cater for different customer needs in a variety of different markets.

C552/045 © IMechE 1998

It is classified as an intermodal system, with intermodal meaning that the freight uses more than one mode during its journey. In this case, it is road and rail, although in other contexts any combination of road, rail, air or sea transit may be classified as intermodal.

Other types of intermodal freight using rail for part of its journey include deep-sea containers or swap bodies on flat rail wagons and Road-Railer. The containers and swap bodies use the same rail wagons; the difference is that containers can be stacked, as they are on ships and on double stack trains in the US, whereas swap bodies can offer the advantages of side loading capability, similar to semi-trailers. They cannot generally be stacked and are often heavier than containers. Both are lifted by crane. Road-Railers are semi-trailers which are strong enough to be coupled together into a train when each is supported by a rail bogie.

The other main type of rail freight system is the conventional wagon. It is confined to rail and therefore has to be loaded and unloaded at a rail connected terminal of factory.

3. FACTORS INFLUENCING CHOICE OF MODE

Freight customers are rarely influenced by anything more than price, how much weight or volume can be carried, and how reliable the service is. They will almost certainly be current road freight and logistics services users, and this will be used as a benchmark against which alternatives will be judged. They will choose whichever system suits them best.

For deep-sea transit, unless the cargo is in bulk, the container is the universal workhorse. It can be carried by rail or road and, in the US, is carried double stack by rail. Double-stack is increasingly used for domestic freight, but the volume sent by piggyback is nonetheless stable at around 3.8 million trailers a year.

Swap bodies are generally confined to rail plus road feeder although a few move by Ro-Ro ferry. Hauliers will not use them for purely road traffics because of the extra weight of a swap body plus the skeletal trailers, compared with just a standard semi-trailers. Road-Railers are more specialist in their use; they are popular for some traffic in the US and are beginning to appear on the continent. They have been trialled in the UK.

Wagonload is generally the most economic between rail-connected premises because it can take higher volumes and more weight than boxes. If these conditions can be met, then the wagon is a very efficient mode in terms of payload and capacity. However, it can have difficulty finding backloads.

Piggyback is less economic than double stack containers but, where these cannot be operated, piggyback is popular and growing. Double stack cannot be operated in Europe because tunnels and bridges are too low. Piggyback offers the lowest tare weight over the road and therefore the highest payload for the haulier.

The Rolling motorway is a drive-on drive-off system used to carry road vehicles across the Alps by rail. It has a high weight penalty for rail, since the tractor unit must be carried and accommodation provided for the drivers, and is at present heavily subsidised.

In making his choice on the basis of the criteria outlined above, the customer will also have regard to his experience with the railways. Many have memories of poor service but, more importantly, some remember making investments in equipment, terminals etc which could only be used by rail. They then remember the railways hiking the prices to make the service totally uneconomic and, in the process, effectively causing the customer to have to write off his investment. This was certainly true in the UK, but is sadly not unique to this country.

Much has changed in the UK and the new operators are seeking to be more aware of customer needs. However, there remains suspicion that will probably take decades to dissipate, causing operators to avoid making heavy investment that can only be used on rail.

Freight customers are becoming more concerned about road congestion and many are now looking at rail to provide a solution to the long trunk leg of many journeys. Since 75% of HGV's on the UK's roads are semi-trailers, it is natural for them to look at piggyback which uses virtually standard lorry semi-trailers, with just the addition of lifting points and other minor modifications which add 5 – 10% to the purchase price of a trailer. The proposals on lorry weights in the recently published Transport White Paper allows the haulier to operate at an extra 3 tonnes gross vehicle weight for a penalty of only 300 kg on the piggyback trailer.

4. PIGGYBACK EXPERIENCE IN THE UK AND CONTINENTAL EUROPE

Piggyback is widely used in the US, and the volumes are large. Several long trains leave major centres nightly for various destinations. The services are generally a mixture of pre-booked and turn-up-and-go, and some operators offer a range of service qualities, from guaranteed delivery to standby. The check-in process usually takes one to two minutes, often without the driver leaving his cab. During this time, the trailer is checked for defects (to avoid future claims for damage) and the driver is given a printed instruction as to where to drop his trailer and where to pick up his return load, if any. The success of these services lies in their appeal to the truckers, where they understand and appreciate the service, the flexibility and the ability to use virtually standard equipment. The service is popular in the US and customers are well used to using it as a simple and attractive alternative to long haul road transport which requires the smallest and least costly change in their operations.

In continental Europe, piggyback services have operated for many years. In certain member states, lorry semi-trailers carry a lower percentage of freight than fixed chassis lorries and trailers which cannot be carried piggyback.

In both the US and Europe, rail wagons are capable of taking either containers, swap bodies or piggyback trailers, giving the rail operators greater flexibility in deployment of his resources to meet changing demand.

5. SQUEEZING THE GREATEST CUBE OUT OF PIGGYBACK

Trailer lengths and widths are limited by national and European regulations and so the customer looking for bulk must have the maximum trailer height. For equipment used on both road and rail, the maximum height on each mode must be considered. For road operation, the maximum height varies throughout Europe but lorries travelling in more than one Member State are, in practice, limited to 4 metres high; to get the most height of load, smaller wheels are often used, although there is a penalty on wear and tear costs.

Trailers operating on piggyback work can be lowered on their suspension by up to 20cm, so the maximum height of trailer required to be carried on rail is 3.8 m. Several designs of piggyback wagons have been produced. All have a common dimension of supporting the trailer wheels 330mm above rail level, the minimum achievable in the UK. Thus, the maximum useful height in the UK for piggyback wagon plus trailer is 4.13 m above rail level (Fig 2). This same dimension is achievable on the continent with GB+ gauge, and makes it a useful standard in that 4 m high trailers in road mode can just be carried on most parts of the continental rail network by piggyback.

Unfortunately, the British rail network cannot at present take such high trailers. The maximum height varies between different routes, but is generally 150 to 300 mm lower. Much of the network cannot even carry 8'6" containers on flat wagons at a time when the deep-sea market is moving rapidly to 9'6" boxes. Of course, low height trailers, tankers and those designed for heavy rather than bulky goods, which may be barrel-shaped and compatible with old railways arches, can be carried, but the heavy ones probably comprise only about 25% of the total market. Similarly, rail transport of containers to and from our ports can take no 9'6" boxes except on special wagons with small wheels or a well between bogies; both have a cost penalty. And 8' 6" boxes can only be handled on some routes. On this basis, the railways are loosing out on a large potential market.

6. HOW MUCH IS EXCLUDED FROM RAIL?

Railtrack is examining a network of rail freight routes linking ports, including the Channel Tunnel, Felixtowe, the Thames, Southampton, Merseyside with London, the Midlands, the North West and Glasgow. These may be high gauge, but how high is not yet clear. If Railtrack accept the Piggyback Consortium's proposal to enhance the Channel Tunnel to Glasgow route to take piggyback, that would attract 400,000 trailers per year off the road if the offer by rail were competitive. If Railtrack does not upgrade this route, the traffic will stay on the road. Similar comments apply to the port traffic, primarily containers but also with the potential for piggyback. Over 2 million trailers a year pass through ports in Britain without drivers (unaccompanied). Many of these could also be attracted to rail, eg London to Ireland via Holyhead, Liverpool, Stranraer, or Heysham. Such door-to-door traffic is now predominantly in 9' 6" containers and could go by rail only if the gauge on these routes were enhanced.

However, with gauge enhancement, the potential amount of traffic which could be attracted to rail is very high, since most long distance road freight is either in boxes or containers. The extent of diversion is dependent on whether the rail system has the capacity, and on the commercial offer, but since it is Government policy to encourage as much freight onto rail as possible, then the gauge upgrade, coupled with capacity enhancement that will be required for all types of rail traffic, can at least provide an acceptable offer to the road freight market.

7. UK GAUGE ENHANCEMENT

The responsibility for maintenance and enhancement of the rail network belongs to Railtrack who have a Licence Obligation to do so and provide capacity for freight ahead of reasonable demand.

The Piggyback Consortium has examined a number of route options for gauge enhancement and, since Railtrack took over responsibility for the infrastructure from British Rail, has generally had positive and fruitful dialogue with the Company. After considering the options, Railtrack decided that the gauge enhancement should be undertaken alongside the West Coast Main line passenger upgrade

Freight operators generally support the use of the WCML for piggyback gauge as well as standard height freight since it serves important terminals and should provide the shortest journey time, an increasingly important issue for time sensitive freight. During the consultations on PUG 2 (Passenger Upgrade 2) on the WCML earlier this year, many operators expressed concern that there would insufficient capacity on the WCML, particularly south of Rugby to carry the expected number of trains requested by Virgin, Silverlink and other passenger operators, as well as the freight operators forecasts outlined above.

At the time of the consultation, Railtrack admitted that they had not produced a timetable for the slow line operation. Timetabling specialists believe that the existing Silverlink service will take up virtually the whole capacity on the slow line once access to the fast line is denied them, unless it is cut back or degraded in various ways. It is not clear that this issue has been properly addressed yet. This leaves little capacity for freight on the slow tracks and none on the fast, if maintenance is all to be done at night.

Although the Consortium proposed alternative routes some time ago, Railtrack decided for operational reasons to use the WCML for piggyback throughout between Willesden and Motherwell. Now it is being suggested by Railtrack executives that WCML capacity is so tight that piggyback gauge could not be properly exploited anyway, depriving Railtrack of a return on its investment.

Whichever route is chosen, Railtrack estimates that it will take about five years to complete the work and open up the route to high gauge traffic. This would not only accommodate piggyback traffic but also 9'6" boxes and high cube wagonload freight, again to the same height as available on the continent with the GB+ gauge.

8. FINANCING THE PIGGYBACK UPGRADE

Railtrack is seeking commitments to use piggyback services from existing and potential freight operators. EWS and Freightliner have presented the following traffic estimates to Railtrack for 2008:

Freight trains (daily in each direction) on part or all of WCML:

	Standard Gauge	High Gauge
EWS	100	20
Freightliner	56	10

In addition, other companies are looking at entering the market, and John Chapman, Deputy Director General of the Road Haulage Association has written to John Prescott 'I am certain that demand for both domestic and international piggyback services will be very substantial if the quality of service and price is right.' Unsurprisingly, companies in the freight industry are unwilling to enter into contractual commitments for a service that will not start until the upgrade is complete in about 5 years' time.

Railtrack has said it intends to seek government financial assistance through Section 139 of the 1993 Railways Act, part as grant and part dependent on the actual traffic generated.

In addition, it would be possible for Railtrack to seek finance from the European Commission under the Trans Europe Network (TEN) Programme and, since the Glasgow – London-Channel Tunnel is a priority route, there are strong indications that Brussels would look on such a funding application sympathetically. We understand that no application has yet been made for such work connected to the freight upgrade.

Of Railtrack's total cost estimate for the work, including disruption, of £250 million, they are expected to seek well over half from Government although, again, at the time of writing, we understand that no application for funding has been made.

9. HAS RAILTRACK GOT ITS SUMS RIGHT?

The original 1994 estimate by the Piggyback Consortium for engineering the upgrade of a route from the Channel Tunnel to Glasgow was well under £100 million. Railtrack confirmed this with several studies, the latest one of which was their Report 'Orbital Route 1b', issued to the Consortium members in November 1997, which gave engineering costs on the same route of £75.2 million (see Table 1).

Table 1 - Cost estimates from Railtrack's Orbital Route 1b Report (1997)			
(£ million)	Engineering	TOC compensation	Total
WCML route to piggyback gauge			
Channel Tunnel – Willesden	30.2	5.2	34.8
Willesden-Northampton-Crewe (slow lines)	16.8	10.4	27.7
Crewe-Mossend	22.1	8.6	30.7
Weaver Junction-Liverpool	4.1	2.3	6.4
Castle Bromwich-Hams Hall	2.0	0.1	2.1
Total	**75.2**	**26.6**	**101.8**
WCML to 9'6" gauge			
Willesden-Rugby-Stafford	2.0	0.6	2.6
Stafford-Mossend	0.4	-	0.4
Total	**2.4**	**0.6**	**3.0**

Since Railtrack has announced that it is going ahead with the 9'6" upgrade, the net cost of piggyback upgrade on the WCML was therefore £98.8 m.

Comparing this with Railtrack's present estimate quoted as £250 m, we understand that this increase may be caused by:

- Railtrack's decision to upgrade all four tracks on the WCML south of Crewe to piggyback gauge. This is estimated to more than double the engineering costs of this section, since this work includes Kilsby and Stowe Hill tunnels south of Rugby, well know to be engineering problems.

- increased disruption costs in payments to train operators delayed by engineering works. These costs were originally estimated by Railtrack as £26.6 m between Crewe and Willesden; they would probably be much more than double this if all four tracks and Kilsby tunnel were upgraded.

- risk of cost over-run. It is understood that Railtrack often add significant additional cost to estimates to cover cost overruns. We have no knowledge of how much.

To summarise, the additional cost of upgrading the WCML fast lines south of Crewe are estimated to be:

	£m	Source
Engineering costs – same as slow line	16.8	Railtrack report:
		Orbital route phase 1b
Compensation to train operators	10.4	ditto
Extra for Kilsby and Stowe Hill Tunnels reconstruction (fast line)		
Engineering costs	40	Piggyback Consortium estimate
Compensation costs	30	ditto
Total	£97.2 million	

If one adds a risk factor of £50 m to these figures, one reaches the £250 m now quoted by Railtrack. It is reasonable to assume that the majority of the risk factor would be attributed to the WCML south of Crewe, and to the fast line, since it is on this section that both the engineering risks and the compensation payments to train operators are likely to be greatest.

From the above comments, it is clear that Railtrack's decision to upgrade all four tracks between London and Crewe to piggyback gauge is responsible for adding at least £100 million to the cost of the project.

The reasons quoted are that the slow lines have to be closed for maintenance at night for one week in three, so the high gauge traffic would have to use the fast lines. North of Crewe, there are four tracks for only sort sections on the WCML to Glasgow; presumably Railtrack has decided on an alternative maintenance regime there which keeps at least one track open – as is normal on main continental railways. South of Crewe, Railtrack has prioritised passenger first, maintenance second and high gauge freight third.

C552/045 © IMechE 1998

10. A SOLUTION TO DELIVER PIGGYBACK GAUGE AND SAVE £100 MILLION

Members of the Piggyback Consortium have come up with a cheaper alternative to upgrading the WCML fast lines between Willesden and Crewe to provide a diversionary route. Railtrack has itself investigated the route Willesden-Reading-Oxford-Leamington Spa-Crewe, reported in their document 'Orbital Route phase 1b'. On this route, upgrading two tracks, engineering costs are quoted as £34.7 m and costs of compensation train operators £6.7, totalling £40.9m (Table 2).

Table 2 – Gauge enhancement on Willesden-Reading-Leamington Spa-Crewe route

Cost estimates for Route **Willesden-Reading-Oxford-Leamington Spa-Crewe** (from Railtrack's Orbital Route Report, Phase 1b)

(£ million)	Engineering	TOC compensation	Total
Piggyback gauge			
Willesden-Oxford-Leamington Spa-Crewe 34.7		6.7	40.9
9' 6" gauge			
Reading-Oxford-Leamington Spa-Crewe 7.8		1.0	8.8

Total additional cost for piggyback gauge on this route - £32.1 m

---

In its 1998 Network Management Statement, Railtrack indicates that it intends to improve the Southampton-Reading-Leamington Spa-Crewe line to 9' 6" gauge in any event, so it appears reasonable to use the addition cost of upgrading this route to piggyback gauge of £32 million.

Table 3 provides a comparison of the costs of creating diversionary routes for when the WCML slow lines are closed for maintenance.

Table 3 - Comparative costs of diversionary routes – Willesden to Crewe

Summary of additional costs to create diversionary route between Willesden and Crewe for use when WCML slow lines are closed:

WCML fast lines – Engineering and TOC compensation	£ 97.2 m
Risk factor – assume 75% of £50 m (see para 9)	£ 37.5 m
WCML total	£ 134.7 m
Reading-Oxford-Leamington Spa	
Engineering and TOC compensation	£32.1 m
Risk factor – assume negligible on this route	-
Saving by using Oxford route	**£102.7 m**

Advantages of the Oxford diversionary route include access to additional terminals on that route, and a widening of the initial piggyback gauge network. Special arrangements would have to be made to ensure access to WCML terminals but, for a saving of over £100m, this does appear worth further study.

Construction of the Oxford route could take place at an early stage, not constrained by the WCML upgrade. Although there are capacity problems on the Reading-London section, this would not be serious since the route would only be used at night.

The Consortium therefore believes that a diversionary route to avoid upgrading the WCML fast lines could be provided for £100 million less than the current Railtrack proposal.

11 THE CASE FOR HIGH GAUGE PIGGYBACK IN THE UK

Piggyback without gauge enhancement to enable 4 m high trailers to be carried will remain a niche market, albeit an important one, suitable for traffic where the greatest volumes are not required. To make any significant impact on road traffic congestion, the full 4 m height capability will be required. The Channel Tunnel-London-Midlands-North West-Scotland route corridor proposed for the initial gauge upgrade follows closely the most congested stretches of motorway in the UK, the M25, M1 and M6 so the argument for upgrading the rail line is strong, especially where it will reduce the need for motorway widening.

Piggyback services are in direct competition with road, where rates are now lower than ever, often well below £1 per mile. To attract traffic onto rail, it has to match the whole journey cost including rail, road haulage at each end and lifts etc at terminals. Historically, in the UK and on the continent, in order to attract and keep intermodal traffic, the railways have often cross-subsidised it from more profitable bulk haulage where the competition is less severe. Reflecting the likely costs and benefits of intermodal traffic, grants are available from the DETR to enable these to be operated competitively with road.

In the longer term, it must be hoped that implementing the provisions of the Transport White Paper (1) and the group of EC papers on the railways issued in Brussels in July 1998 (2) will enable intermodal freight to compete more fairly with freight using road all the way. But this requires work to enable the railways and their owners to identify and understand their costs, and the elements applicable to rail freight and to become more efficient.

12 PIGGYBACK SERVICES IN THE UK

Piggyback services, using low height trailers, have started nightly between Glasgow and London (EWS), and Manchester and Tilbury (Freightliner). New wagons from Thrall and Babcock, which can carry both containers, swap bodies and trailers with full width wheels are on order, and these and other operators are considering expanding their services both in frequency and destinations served.

These low height services can operate to most locations in the UK and are likely, at least to start with, to be developed with particular customers in mind as part of existing services.

With higher gauge clearances, the potential is dramatically increased. Cross-Channel services from destinations between London and Glasgow can operate to virtually any continental town connected to the standard gauge network. Shorter hauls being examined include the North West/Midlands to Lille as well as Glasgow to London, Glasgow to Manchester and Manchester to London. The shorter hauls might eventually support services as frequent as hourly, thus providing a genuine turn-up-and-go service with minimum wait as an alternative to the adjacent motorway.

13 RAILTRACK'S OBLIGATIONS TO FREIGHT

Railtrack has made a clear commitment, enforceable by the Regulator, to provide capacity ahead of reasonable demand for freight anywhere on the network. *'Never again will freight be turned away'* Government Response to the House of Commons ETR Committee's Report 'A New Deal for the Railways', July 1998

The demand for piggyback has not changed for several years. It must be recognised that freight cannot give firm, risk-free commitments to Railtrack as passenger TOCs can. The forecasts by EWS and Freightliner, coupled with positive statements by the Road Haulage Association are very much firmer than road haulage companies would give to the Highways Agency assessing whether a new road should be built.

The Consortium believes that Railtrack has an obligation to provide the capacity for freight, including high gauge traffic, only subject to satisfactory funding arrangement. If piggyback gauge is ruled out of the WCML due to the urgency of finalising contracts for the passenger upgrade, that is entirely due to Railtrack not developing the project early enough or fast enough to enable its Board and the DETR to negotiate a funding package in time.

An alternative reason may be that Railtrack has now concluded that there will not be enough capacity, even after upgrade and the eventual introduction of moving block signalling. However, it is evident from many sources within Railtrack that piggyback gauge work on the West Coast Main Line must go ahead by the end of the year to fit in with the WCML passenger upgrade.

Whatever the route, it is in any event time to stop the studies and get down to action on the ground. Railtrack's current estimate of the construction time to create the gauge enhancement is still five years.

14 PIGGYBACK'S SUPPORTERS

Government documents and statements by John Prescott indicate strong support for this project, which has the potential to attract at least 400,000 lorries per year off the motorways between Glasgow and the Channel Tunnel and many more within the UK. The two main freight operators, EWS and Freightliner, say that the higher gauge will generate at least 30 more trains per day in each direction on the west Coast Main Line.

The Road Haulage Association and freight logistics providers have confirmed strong interest among potential users.

Other relevant comments include:

'.we have taken the lead in developing and investing in the (piggyback) project...We have now finalised the previous estimates of the capital costs for enhancing the route via London and the WCML...The next step is for us to work in parallel with the DETR, ORR and our customers to finalise a funding package. From our discussions to date, we expect part of the funding to be provided by the DETR under its existing mechanisms for encouraging the transfer of freight from road to rail.' Railtrack's Network Management Statement, February 1998.

We will also look at the contribution that intermodal freight terminals and 'piggyback' style operations...could make to increasing rail freight's share of the market....We will consider applications from Railtrack and others for additional public investment on a case by case basis'. White Paper 'A new deal for transport: Better for Everyone' DETR, July 1998

Rail freight is a better environmental option and we endorse the rail freight industry's target to double traffic carried by rail within 5 years and treble it in 10'... 'It is my intention to use the proceeds from the part-sale of the NATS (National Air Traffic Services) to reinvest in transport improvements, such as a Euro-gauge freight rail system from the Channel Tunnel right through to Scotland if terms can be agreed.' Deputy Prime Minister John Prescott speaking in the House of Commons 20 July 1998

If Railtrack fails to deliver on addressing (its licence obligations in respect of freight) then it will be liable to enforcement action by the Regulator. Michael Beswick, ORR

'To derive our base demand forecasts, we have made the following assumptions (among which is) no fundamental change in the industry's structure, its current equipment or operating formats apart from further growth in intermodal options, including the introduction of various forms of piggyback trailer' Railtrack's Network Management Statement, February 1998

And in the future, *'The Strategic Rail Authority will be the Government's principal agent in ensuring that freight is never again turned away from the Rail Network'* Govt. Response to House of Commons ETR Committee's Report 'A New Deal for the Railways', July 1998

C552/045 © IMechE 1998

References

1. Government Transport White Paper: 'A New Deal for Transport: better for everyone', July 1998 HMSO Cmnd 3950

2. European Commission Papers 98/0265 (COD), 98/0266 (SYN) and 98/0267 (SYN) comprising:

- Proposal for a Council Directive to amend Directive 91/440 on the development of the Community's railways.

- Proposal for a Council Directive to amend Directive 95/18 on the licensing of railway undertakings.

- Proposal for a Council Directive relating to the allocation of rail infrastructure capacity and the levying of charges for the use of railway infrastructure and safety certification.

- Commission Working Paper – explanation of the individual articles in the proposal for a directive relating to the allocation of railway infrastructure capacity and the levying of charges for the used of railways infrastructure ad safety certification.

Appendix

List of Piggyback Consortium Members

Abbcott Estates Ltd
Angel Trains Ltd
Babcock Rail
Bombardier Portal
CAIB UK
Cartwright Group
Central Sea Corridor Joint Task Force
Combined Transport Ltd
Daventry International Rail Freight Terminal
Dublin Chamber of Commerce
Dumfries and Galloway Regional Council
English Welsh and Scottish Railway
European Intermodal Association
Eurotunnel
Freight Transport Association
Freightliner Ltd
Isle of Anglesey CC
John Russell Ltd
Kent County Council
Local Government Association

MDS Transmodal
Mersey Docks and Harbour Co Ltd
Merseyside Authorities
North West Channel Tunnel Group
North West Regional Association
Novatrans
Rauteruukki
P & O Ferrymasters GmbH
Powell Duffryn Rail Projects
Powergen Property
Railfreight Distribution
Rail Freight Group
Railtrack (until July 1998)
Servant Transport Consultants
Scottish Enterprise
SERPLAN
SNCF Fret
Surrey County Council
Thrall Car Company
Transport 2000
UIRR

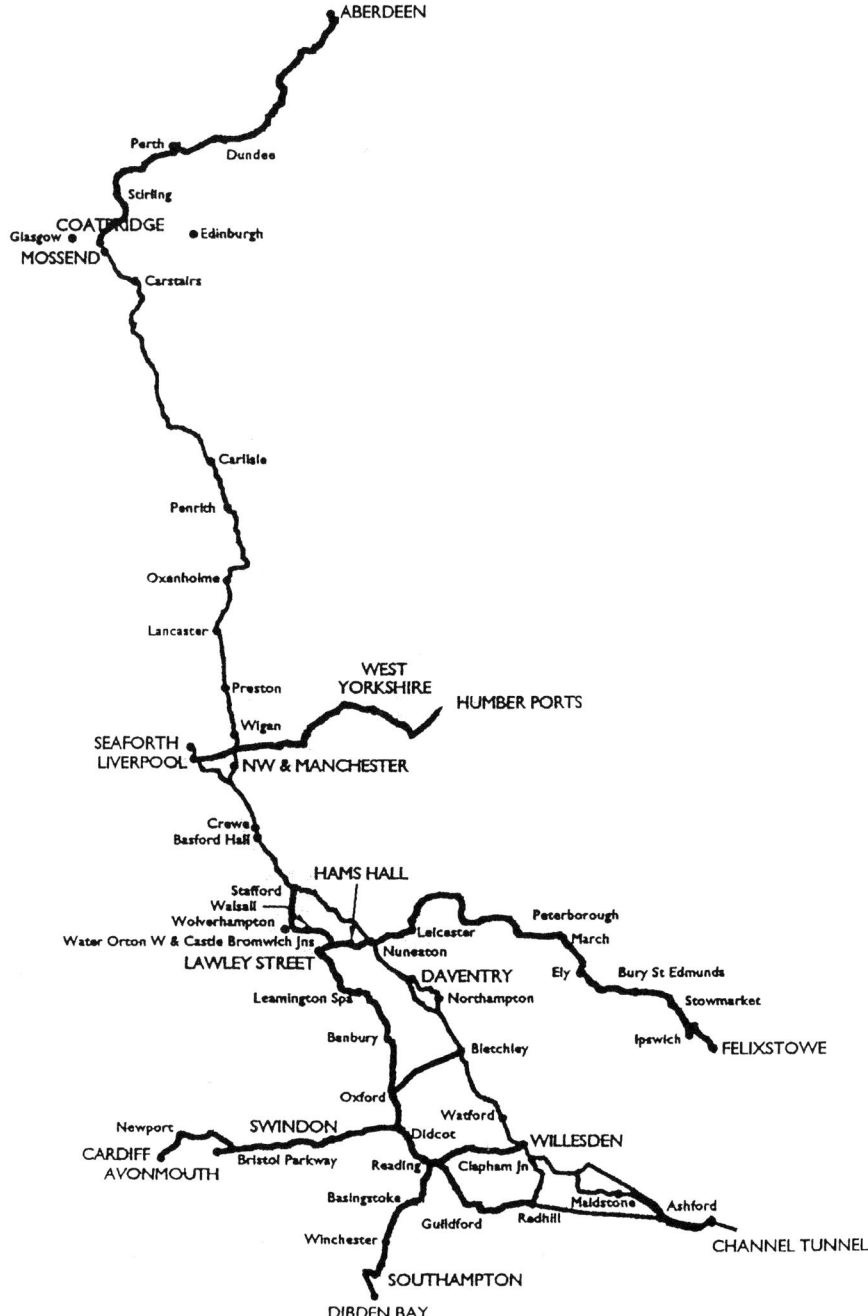

Fig 1. Proposed high gauge routes – source Railtrack

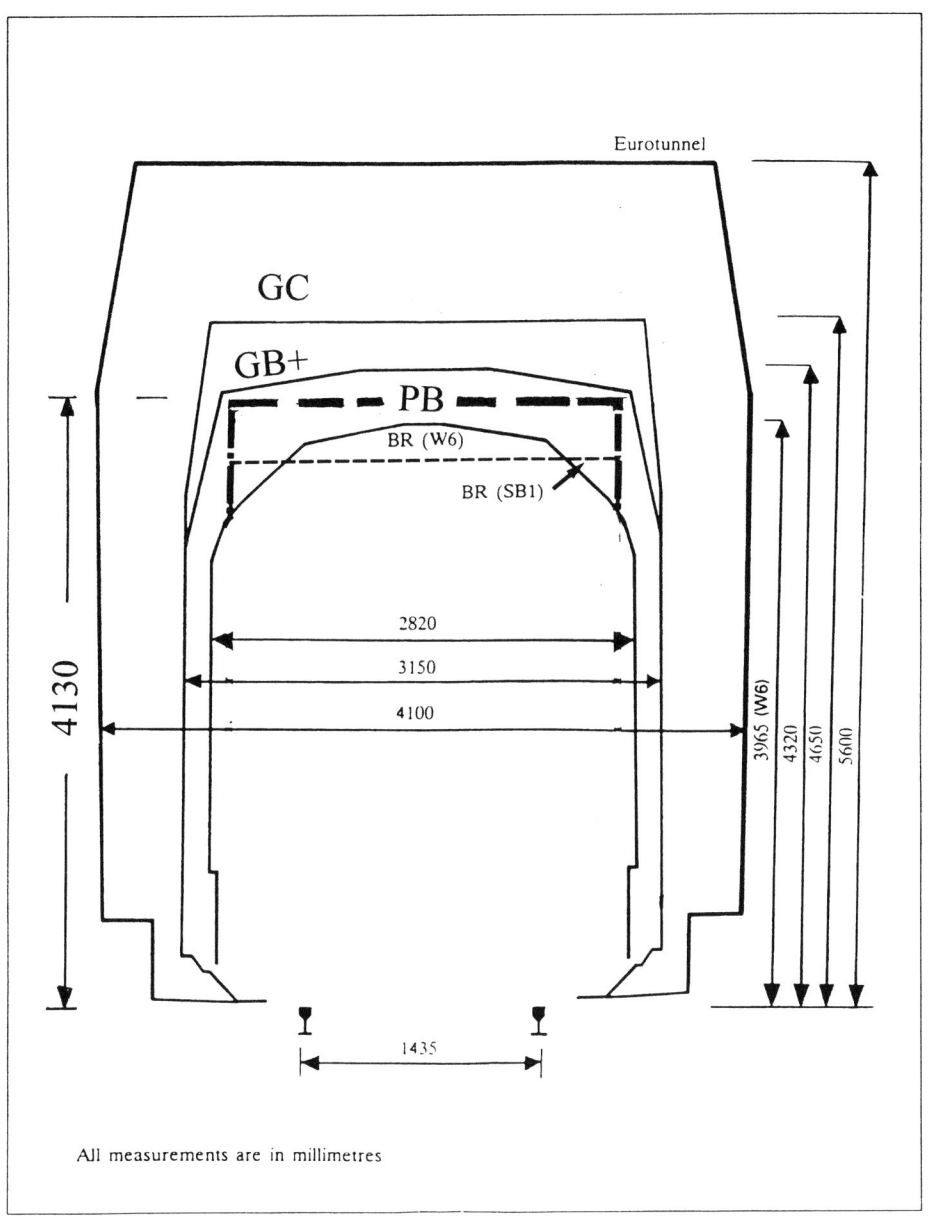

Fig 2. Rail structure and loading gauges

The Thrall Europa Eurospine wagon at the launch of the service at Deanside, Glasgow.

C552/042/98

The bimodal vehicle as a technology to grow rail freight

B T SCALES BSc, PhD, MASME, FIMechE
Transportation Engineer, Pittsburgh, USA

Bimodal vehicles capable of operating on both the highway and the railway could be the optimum technology for transportation of merchandise freight in future. This paper reviews existing technology, suggests requirements for bimodal vehicles to suit the unique conditions of the United Kingdom and proposes a conceptual bimodal vehicle technology to meet this challenge.

1.   INTRODUCTION

Technical papers are expected to be concerned with the past. They describe design, testing, research and development of systems and/or items that have been completed. In most cases, this work has been implemented, with the systems and/or items described being in use at the time of presentation. This paper is different. It is concerned with the future. It proposes an innovative way ahead for rail freight in the United Kingdom for the Twenty-first Century, namely the development of self-contained bimodal vehicles capable of operating on both the highway and the railway systems.

This paper includes the following items:

- Review of existing bimodal vehicle concepts.
- Design constraints for British bimodal vehicle.
- Requirements for successful British bimodal vehicle.
- Description of proposed "Chameleon" bimodal vehicle.
- Advantages of proposed "Chameleon" bimodal vehicle system.
- How bimodal vehicles can switch freight from road to rail.

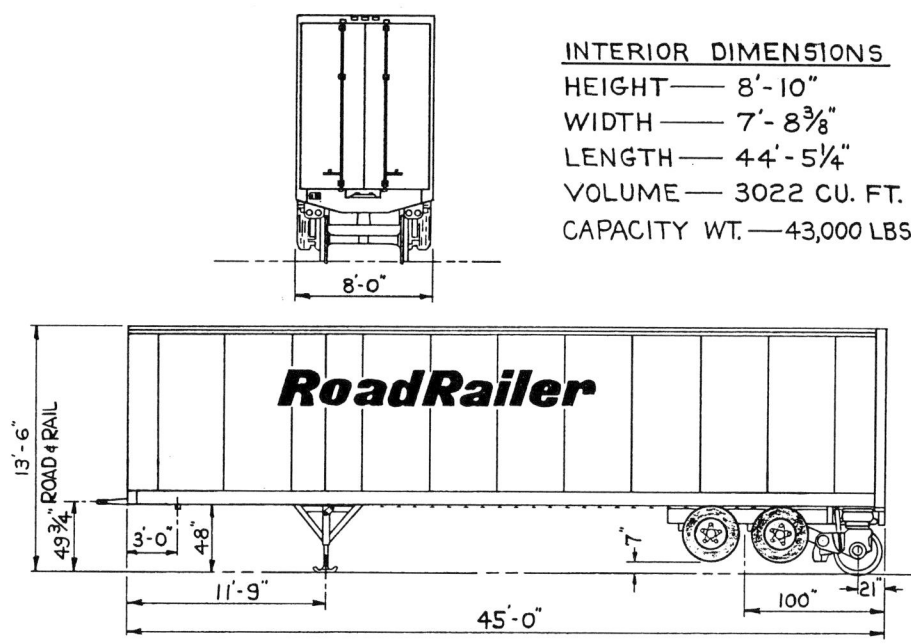

INTERIOR DIMENSIONS
HEIGHT —— 8'-10"
WIDTH —— 7'-8⅜"
LENGTH —— 44'-5¼"
VOLUME —— 3022 CU. FT.
CAPACITY WT. —— 43,000 LBS.

Fig. 1 RoadRailer MK. III

Fig. 2 RoadRailer MK.V

C552/042 © IMechE 1998

2. REVIEW OF EXISTING BIMODAL VEHICLE CONCEPTS

A number of bimodal vehicle concepts have been proposed in the past. Prototypes have been built and demonstrated in the U.S.A. and Europe at various times. The only concept that has been put into regular service is the Roadrailer, which has been developed through several "Marks" from the original version that was designed by the Research Department of the Chesapeake and Ohio Railway in the U.S.A. about fifty years ago. (1)

The first Roadrailer consisted of a "pup" trailer, equipped with a single highway axle and a railway wheelset at the trailing end, each fitted with air suspension. The leading end of the trailer incorporated a tongue that could engage and lock with a socket in the rear of the vehicle ahead when in the rail mode. The change-over between road and rail was effected by activating the appropriate air suspension. The change-over took place on a siding that was paved up to rail top level. An adapter vehicle was required to connect a rake of Roadrailers to a locomotive or an existing train. This adapter vehicle was equipped with a conventional coupler at one end and a socket for the tongue of the first Roadrailer at the other end. The original Roadrailers carried mail and ran at the rear of passenger trains for some years until the passenger train service was withdrawn.

The next stage in Roadrailer development was the "Mark III", shown in Fig. 1, which consisted of a modified full-size highway trailer, with the addition of a single railway wheelset at the trailing end. A 50-vehicle train of this type was demonstrated in the United Kingdom in the early 1960's.

The final stage of development is the "Mark V" shown in Fig. 2, where the permanently attached railway wheelset has been replaced by a detachable modified 3-piece bogie, which supports the trailing end of one Roadrailer and the leading end of the next vehicle.

The requirement for an adapter vehicle for every train and the need for an adequate supply of spare bogies for the "Mark V" version are hurdles to be overcome in the implementation of this Roadrailer concept. An additional disadvantage in the case of the United Kingdom is that the Roadrailer V is approximately 10 cm. higher in the rail mode than in the road mode. This situation is unfortunate because rail vertical clearances are more restrictive than highway limitations for the design of a bimodal vehicle to be used in the United Kingdom.

Roadrailer vehicles currently operate the following services in the U.S.A.:

- Transporting automobile components in block trains.
- Carrying mail at the rear of passenger trains.

3. DESIGN CONSTRAINTS FOR BRITISH BIMODAL VEHICLE

The design of a bimodal vehicle is constrained by both highway regulations and railway requirements, with the more restrictive case controlling. In North America, highway regulations control every aspect. In the United Kingdom, vehicle height is limited by the railway loading gauge. Highway regulations control other features such as length, width and gross weight.

At the present time, potential routes for innovative intermodal systems are being reviewed with the objective of designating a number of trunk rail routes with increased overhead clearance in order to improve the viability of these innovative concepts, such as bimodal vehicles, piggyback arrangements and hi-cube containers.

Allowable gross weight for a bimodal vehicle is influenced by the recent change in highway regulations that permits an increase from 38 tonnes to 44 tonnes for an 18-wheeler

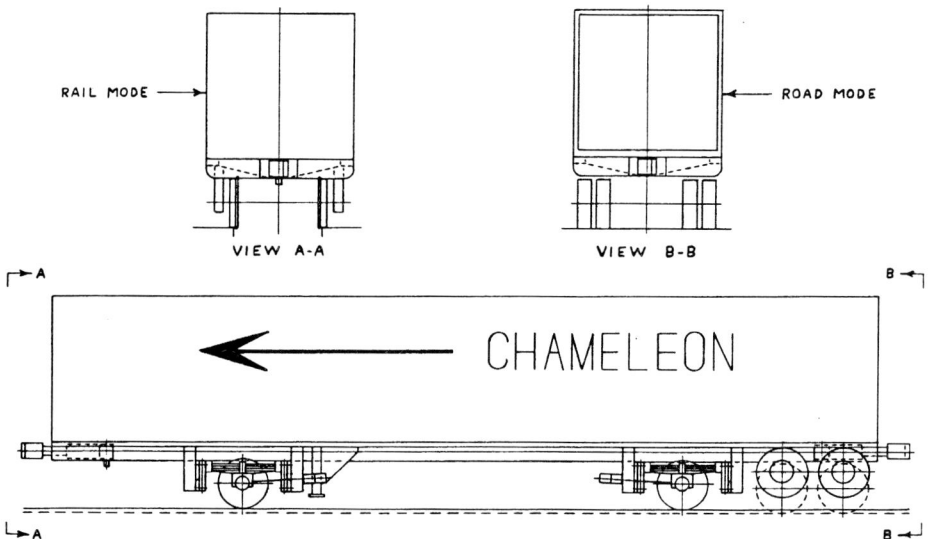

Fig. 3 Rail-Road Box Car

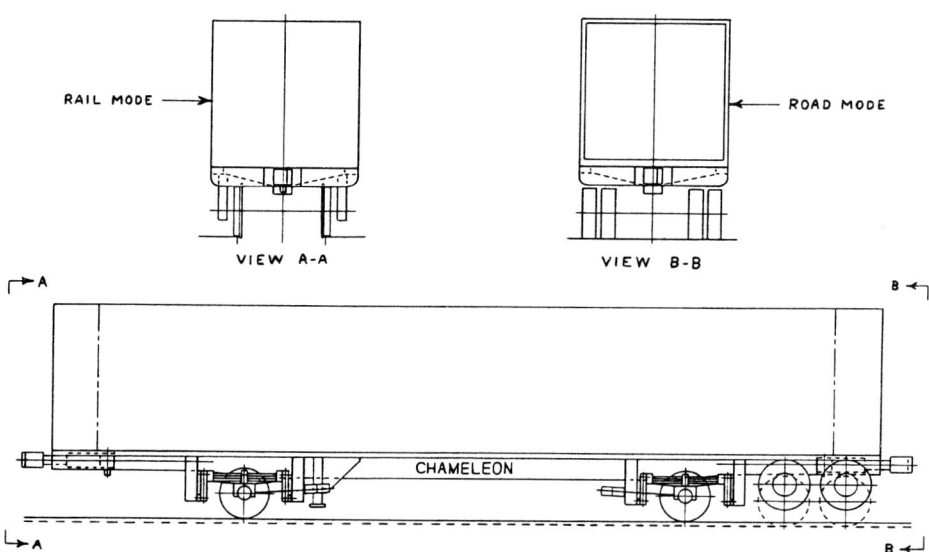

Fig 4 Container Carrier

C552/042 © IMechE 1998

tractor-trailer combination during the road-haul portion of an intermodal road-and-rail movement. The direct consequence of this increase is that the additional weight of the rail equipment of a bimodal vehicle does not result in a loss of carrying capacity compared with an all-highway equivalent trailer. It is reasonable to assume that even a conservative design of the necessary rail equipment will add less than 6 tonnes to the weight of the basic highway trailer configuration.

4. REQUIREMENTS FOR SUCCESSFUL BRITISH BIMODAL VEHICLE

The following requirements for a successful bimodal vehicle are suggested:
- Totally self-contained, no 'adapters' permitted.
- Fully compatible with the existing rail system.
- Bi-directional in the rail mode.
- Able to survive in the rail environment.
- Acceptable for limited operation on the highway.
- Container carrying version possible.

5. DESCRIPTION OF PROPOSED "CHAMELEON" BIMODAL VEHICLE

5.1 General Description
The Chameleon bimodal vehicle is a two-axle rail car, capable of running on the highway supported by a retractable road bogie at the rear in conjunction with a highway standard tractor incorporating a fifth wheel for conventional semi-trailers. The Chameleon vehicle was originally designed as a box car of 28 tonnes capacity. A flat car version to carry containers has also been designed using the same components. Unlike other bimodal vehicles, the Chameleon vehicle does not require adapter units to make the change-over from road to rail. It is completely self-contained and fully compatible with the existing rail system. The change-over between road and rail requires only a flat yard with paving up to the top of the rails and a compressed air power supply. The Chameleon intermodal system will be described under the sub-headings of Design and Terminal.

5.2 Design
The Chameleon vehicle, shown in Figure 3, consists of a skeletal frame supporting a lightweight body, with two rail wheelsets and a retractable road bogie containing two axles and low profile wheels. The frame incorporates a square section center sill and lateral bolsters connecting with the suspension assemblies of the rail wheelsets. The ends of the center sill are flared in order to accommodate hydraulic cushion units. Conventional knuckle couplers are provided at each end. Standard road and UIC rail air brakes are installed. The rail suspension assemblies consist of two-stage leaf springs for the vertical suspension and swing links for the lateral suspension with controlled friction damping.

The unique feature of the Chameleon vehicle is the retractable road bogie. Air-operated screw jacks raise or lower the road bogie to make the change-over between rail and road operation. The road bogie is positively locked in both the extended (highway) and retracted (rail) positions. Air-operated jacks adjacent to the leading end rail suspension assemblies are used as landing legs for highway operation. The jacks are also used in the change-over between road and rail, to be described in the next subsection.

Road Mode

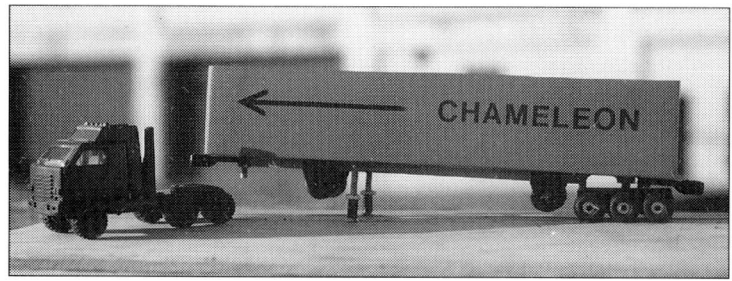

Changeover

Rail Mode

Fig. 5 Rail-Road Vehicle Modes & Changeover

The body is fitted with conventional end doors for use with a highway-type loading dock. The kingpin for the fifth wheel of the highway tractor is mounted on the underside of the center sill. It is, therefore, lower than the standard height. For this reason, the front of the Chameleon vehicle is slightly higher than the rear when running on the highway.

The flat car version shown in Figure 4 uses a fish-belly type skeletal frame and suspension components as before. The frame is equipped with twist locks to locate a 45-foot container, a 40-foot ISO container, or a demountable body.

5.2 Terminal

A yard with paving up to rail level is required for the Chameleon vehicle to make the change-over from road to rail, which is achieved in two steps, as shown in Figure 5. A compressed air supply is needed to make the change-over. The compressed air may be supplied from a hostler tractor or a ground line.

The change-over is made with the Chameleon vehicle parked over the rails, initially supported on its road wheels at the rear and its jacks towards the front, the jacks serving as landing legs. In the first step, the road wheels are retracted. The jacks are released in the second step, positioning the Chameleon vehicle on the rails, supported by the rail wheels. The Chameleon vehicle can now be coupled with other Chameleon vehicles to form a train. The change-over from rail to road is made in the reverse order, finishing with the Chameleon vehicle supported on its road wheels at the rear and its jacks towards the front, ready for pick-up by a highway tractor.

In other respects, the terminal is similar to other intermodal terminals, having a terminal office, inbound and outbound gates and a storage area for Chameleon vehicles in the highway mode.

6. ADVANTAGES OF PROPOSED CHAMELEON BIMODAL VEHICLE SYSTEM

6.1 Comparison with other bimodal vehicles

In addition to combining the best features of road and rail transportation while maintaining complete compatibility with each system, the Chameleon bimodal vehicle is entirely self-contained. Other bimodal vehicles require adapter units, which are a potential source of trouble. Change-over between road and rail and coupling of vehicles in the rail mode are easier than for other bimodal vehicles. A simple terminal is sufficient. Storage space for adapter units is not required.

6.2 Comparison with other intermodal systems

The Chameleon concept as a flat car permits the intermodal transportation of containers to any terminal which has paving up to the top of the rails. Lifting equipment is not required for the containers to change between road and rail transportation. Rail cars are not required for the rail mode and chassis are not required for the road mode, resulting in lower capital costs.

According to a study carried out some years ago in the U.S.A., shown in the form of a bar graph in Figure 6, linehaul costs for bimodal vehicles can be expected to be 25% less than for the equivalent container-on-flat-car operation. (2) This comparison is based on Bar D versus Bar F in Figure 6. It would be reasonable to assume the same order of saving in the United Kingdom. Another study in the U.S.A. compared the characteristics of the proposed Chameleon bimodal vehicle system with the existing trailer-on-flat-car system. This study predicted a saving of 26% in fuel consumption for the proposed system. A saving in fuel

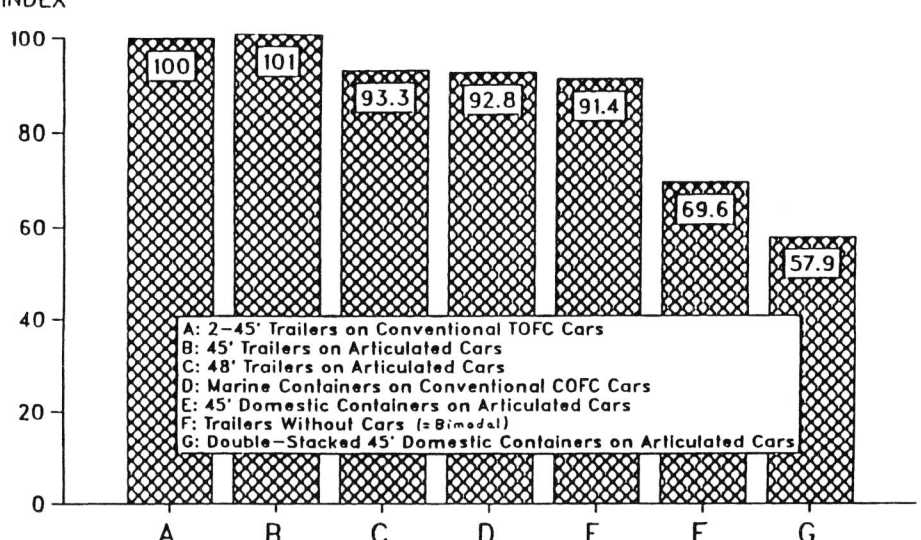

ESTIMATED LINEHAUL COST* COMPARISON
FOR VARIOUS INTERMODAL CONFIGURATIONS
(INDEX COSTS ON A PER TRAILER OR CONTAINER BASIS)

A: 2-45' Trailers on Conventional TOFC Cars
B: 45' Trailers on Articulated Cars
C: 48' Trailers on Articulated Cars
D: Marine Containers on Conventional COFC Cars
E: 45' Domestic Containers on Articulated Cars
F: Trailers Without Cars (= Bimodal)
G: Double-Stacked 45' Domestic Containers on Articulated Cars

* Excludes Capital Costs of Locomotives, Railcars, Trailers and Containers
Source: AAR and TT—BMP Analysis

Fig. 6

consumption indicates a saving in motive power costs also. An existing locomotive could haul 35% more vehicles or a smaller locomotive would be suitable for an equivalent train of Chameleon bimodal vehicles. Similar savings are to be expected in the United Kingdom.

New terminals for Chameleon bimodal vehicles can be built easily and at low cost because lifting equipment is not required.

Compared with proposed British piggyback trailer-on-flat-car configurations, the bimodal vehicle permits a significantly greater cubic capacity, which can be more important than weight capacity in some cases.

7. HOW BIMODAL VEHICLES CAN SWITCH FREIGHT FROM ROAD TO RAIL

7.1 Private rail sidings re-created

In olden times, most industrial facilities possessed a rail siding that gave direct access to the railway network so that both inbound and outbound shipments could be made by rail. Today, only a few such industrial sidings remain, resulting in freight shipments being made by road usually. The bimodal vehicle can effectively re-create the private siding if simple terminals are provided at reasonable intervals on trunk rail routes. With such terminals available, bimodal vehicles can be collected up from and returned to the nearest terminal with only a short distance travelled by road.

7.2 Designed to meet customers' requirements

In the UIC Rail Plan, "Making Rail the First Choice", it is pointed out that bimodal technologies can be tailored to customers' specific needs. (3)

7.3 Service reliability

Delivery times anywhere in the country could be guaranteed with an established and time-tabled railway service for bimodal vehicles in place. In a country the size of the United Kingdom, overnight delivery almost everywhere should be achievable. Freight transportation could become almost as simple and reliable as the British Postal Service, which is what the customer demands and expects.

7.4 Environmental aspects

The greatly reduced rolling resistance of the steel wheel on the steel rail compared with the rubber-tyred wheel on the highway results in a major reduction in energy consumption for the rail portion of an intermodal shipment compared with an over-the-road shipment. This reduction is environmentally friendly as well as economically advantageous for the customer.

Transportation by rail is very much safer than by road. In the U.S.A., for example, heavy highway freight vehicles are involved in a disproportionately large number of road accidents. This aspect alone should be taken more seriously than it actually is.

7.5 Public policy

It is stated government policy to encourage freight transportation by rail rather than by road. The bimodal vehicle facilitates this policy because the advantages of a highway semi-trailer are retained while providing the benefits of rail transportation for the trunk haul. Governmental blessing and assistance are to be expected in view of the anticipated benefits accruing from the introduction of freight transportation by bimodal vehicles.

8. RECOMMENDATIONS

1. Detailed designs of a Chameleon bimodal van (box car) and a bimodal flat car (container carrier) should be undertaken in accordance with specifications to be agreed with potential operators and appropriate government departments.

2. Economic evaluations of the above designs should be undertaken to justify the construction and testing of prototype vehicles.

3. Assuming positive results are achieved from implementation of the above recommendations, a demonstration service should be instituted in order to establish the actual benefits achievable by widespread adoption of bimodal vehicles in the long term.

9. CONCLUSION

The past, present and future of bimodal vehicles have been described. Bimodal vehicles designed to suit the unique conditions prevailing in the United Kingdom have the potential to grow rail freight for the following reasons:

- Economic advantages should be achieved.
- Private rail sidings are re-created effectively.
- The concept is environmentally friendly.
- Safety is enhanced.
- Public policy should be encouraging.

REFERENCES

1. Roadrailer vehicles now supplied by Wabash National Corp.
2. "A Study of the Current Surface Transport Inter-relationships Affecting TOFC Growth", Transportation Research and Marketing, TTX Company, Chicago, IL, 1984.
3. "Making Rail the First Choice", International Union of Railways, UIC Communications Dept., Paris, 1997.

C552/049/98

INNOCAR – an innovative container carrier

G J BAZUIN MSc
N S Materieel Engineering, Utrecht, The Netherlands

Synopsis
Rail freight transport has to be revitalised in order to compete with the other modalities. In the past years less attention has been paid to innovate the existing rail system. By applying proven technologies, like information technology, the rail modality could become more cost effective, more flexible and even more friendly towards its environment.

INNOCAR offers a concept of dedicated rolling stock to favour the transport of large volumes of containers on the European Freight Freeways.

1 Future of rail bound freight transport in Europe

The transport of freight within the European borders is growing strongly. Unfortunately the railways did not profit as much from this as the other transport modalities did. In the past 25 years the rail contribution decreased from about 30 to 15 % (1,2). In absolute figures one has to conclude that the volume did not decrease, from which one has to conclude that the other modalities have gained from the growth.

Current rail freight transport has some disadvantages regarding the gross transport time (from door to door), an unfavourable price - quality balance and a high order of inflexibility. Some specific advantages like the possibilities of introducing a high grade of automation, having at least an image of being environmental friendly and the relatively high standard of safety are exploited insufficiently (3).

In order to revitalise the position of European rail freight transport the transport of large volumes has to become more efficient. To accomplish this several improvements have to be made. An important step is the introduction of the so called European Freight Freeways, which in fact form a European network of railway lines dedicated to freight transport. The first new build Freight Freeway, dedicated to freight transport, in Europe will be the Betuweroute in the Netherlands. This line connects the Rotterdam harbour with the East and South European countries. The Betuweroute will come into service in 2005.

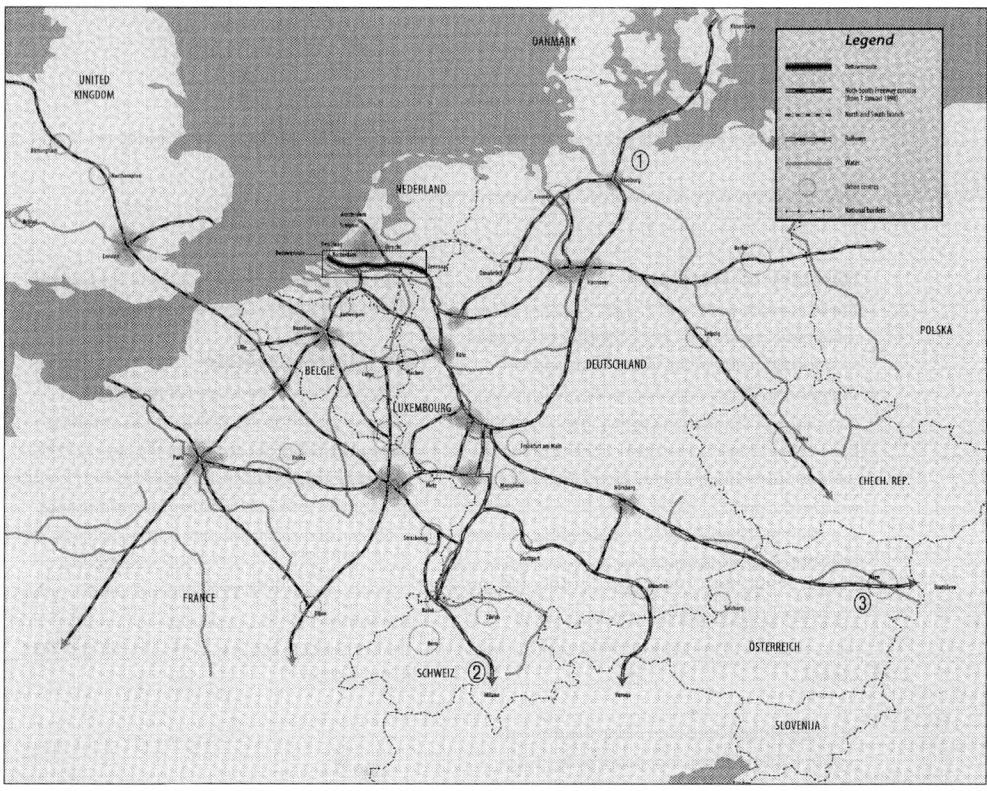

The Betuweroute connecting Rotterdam harbour with Eastern and Southern Europe

From the idea of having dedicated infrastructure the Managementgroep Betuweroute, on behalf of the Dutch government, commissioned NS Materieel Engineering to develop a rolling stock design that is best suited for the transport of large volumes of containers.

2 Development goals

In consultation with the Managementgroep Betuweroute three main development goals were defined:
- a raised efficiency;
- a lower noise emission;
- an innovative image.

These topics will be discussed in the next paragraphs.

2.1 Efficiency

The efficiency of rail bound freight transport has to be improved. Based on a SWOT analyses (4,5) a list of critical points was obtained which has been used as a guideline in the development process. A selection of the most important and promising ideas is listed below:

- shortening the gross transport time by optimising the following processes:
  - loading and unloading;
  - marshalling and preparation of the trains;
  - technical inspections;
  - the transport itself.
- enlarging the payload;
- making more profit from the allowed train length and dimensions (gauge);
- eliminating the manual activities as much as possible in order to lower personnel costs.

2.2 Low noise emission

Trains generally enjoy a fairly positive environmental image. Freight cars, however, are notorious for their noise emission during running. On an European scale it has been agreed that noise emission standards for rail vehicles will be introduced within a few years. Such standards have been in use for several years for automobiles, vacuum cleaners and a broad range of other products. Although most European countries have regulated the noise immission along railway lines, only a few countries, like Austria, have set emission standards that should be fulfilled to get the rail vehicles accepted (6).

In the Netherlands Railned, the body that is in charge of capacity management, is confronted with the fact that the capacity of certain lines is constrained by the noise emission. From this experience and the expectation that noise emission of freight cars will be bound to European standards in shortly it was clear that a new car design should be a low noise design.

2.3 Innovative challenge

The design of freight cars has not substantially altered during the last 50 to 100 years. In the meantime several large technical breakthroughs have been accomplished. Especially in the field of information technology, low noise technology and light weight constructions big progress has been made. The INNOCAR should bring together technologies which have been proven in other applications but are new in the field of the railways. By bringing these technologies together in one prototype train operators and fleet owners as well as the railway industry should be forced to think about which innovations are of interest for them.

3 The INNOCAR concept

The INNOCAR concept is in fact not just a single wagon but a set of four coupled units. These units are coupled by articulated bogies and a coupling rod in the middle (see sketch below).

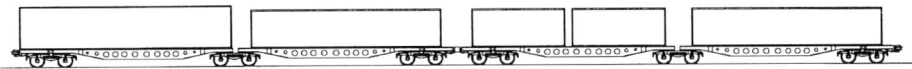

The INNOCAR concept

Each unit is set to accommodate 45' (13.8 m) long containers as it was notified that these containers show a growing market share.

3.1 Loading possibilities

The loading possibilities of the INNOCAR, based on their dimensions, are shown in the following table below.

Compatible loading units

ISO containers	2 x 20' ISO	
	1 x 30' ISO	
	1 x 40' ISO	
	1 x 45' ISO	*high cube* (9.5')
Swap bodies - symmetric	groups: 20 - 24 and 26	on 20' fastening
(according to UIC 592-4)	groups: 30 and 31	on 30' fastening
	groups: 40, 42, 44 and 45	on 40' fastening
Swap bodies - asymmetric	groups: 81, 82, 84, 85 and 86	on 30' fastening
(according to UIC-592-4)	groups: 91,94,96 and 97	on 40' fastening
	MEGA-300 and JUMBO	on 40' fastening

Besides the dimensions of the containers and swap bodies the gross weight is nowadays an important criteria for the loading of a wagon. For example a 2-axle flat car can accommodate two 20' containers in regard to their dimensions, but when these containers are maximal loaded (gross weight 30,480 kg) it is from the maximum axle load of 22,500 kg not allowed to place more than one container on this wagon. This evaluation on weight makes the logistics during loading very complicated and thus time consuming. Therefore it was decided to configure INNOCAR in such a way that every loading unit that can be placed on the wagon from its dimensions could be loaded without the need to check for the axle load. This is effectuated by applying effectively three wheelsets per wagon unit. As there is a clear tendency towards the admission of 25,000 kg axle load (7) the maximum gross weight of the wagon is restricted to 75,000 kg. Having a tare wagon weight below 14,000 kg the possibility is created to accommodate two maximal loaded 20' containers.

Another way to ease the loading process is found in the foreseen development of self adjusting container fixing points. Nowadays a marshaller has to walk along the train at both sides to set the fixing points in the right configuration. For a freight train of 700 m length this activity will take almost half an hour. INNOCAR offers a system in which the container or swap body creates its own fixation points by pushing away the unneeded points. A signalling system indicates to the train driver whether or not the fixation is correct.

By setting the load floor height at 1020 mm even the highest swap body (height 3200 mm) can be transported within the UIC - GC loading gauge. This gauge is defined as the gauge that has to be fulfilled when building new infrastructure or adapting existing infrastructure. However it will take years before the first line offering this GC-gauge comes into service. Until then the maximum height of the loading units is 2,900 mm for these lines which offer the raised loading gauge according to UIC - 505-1. This gauge is the standard in, for instance, the Netherlands, Germany and the main lines in Switzerland. In cases where only the standard UIC loading gauge is available the height of the loading unit has to be limited at 2,595 mm.

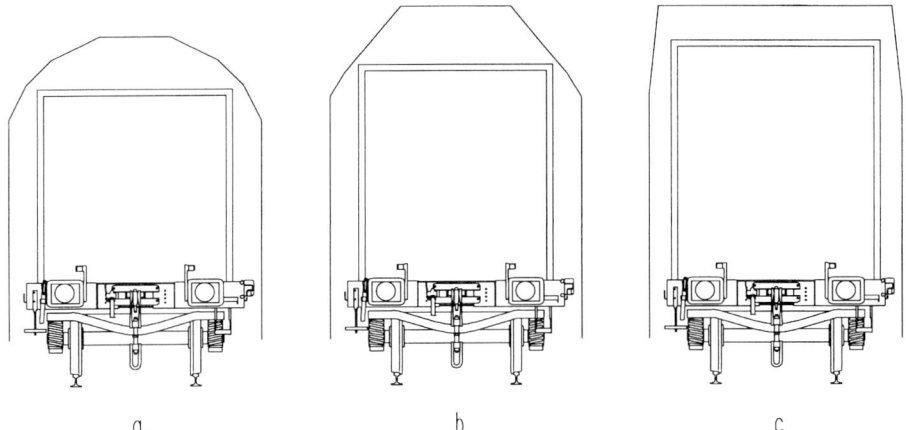

Relation with the allowed loading gauge
a. UIC 505-1 standard (container height 8,5')
b. UIC 505-1 enlarged (container height 9,5')
c. UIC 506 GC (highest swap body 3200 mm)

3.2   Wagon handling

The wagon handling is eased by applying automatic couplers, having wagonsets instead of individual wagons and having automated processes like the functional test of the braking system.

To automate processes like the braking test it is necessary to have information transfer between the wagon units and the locomotive, where the driver is located. This is accomplished by applying train bus technology. Having such an infrastructure for information exchange other functionalities can be added like:

- diagnosis; by applying sensors information can be obtained about, for instance, the wear of the brake blocks, the roundness of the wheels, the temperature of the wheelset bearings and the weight of the load (should remain constant during transport ...).
- electronically controlled brake system which has the advantage above the conventional pneumatically controlled system that the reaction time of all brakes in the train will be extremely short. For a conventionally braked freight train with a length of 700 m it takes more than two seconds before the last wagon notices the brake command!

3.3   Tracing & tracking

The aim of tracing and tracking is to improve the train operator's logistical process and the provision of information to clients. By using for instance GPS and GSM Rail it has become possible for the INNOCAR to determine its position and to send this information to various wall board authorities. Coupled to the information of the location other information in regard to the actual status of the load, the wagon condition etc. can be exchanged. For NSM Engineering it is clear that one has to define a standard for this application on, at least, a European scale. This statement also counts for the choice of the train bus. Several projects on these topics have recently started or will start in the near future (1).

3.4 Low noise design

The production of rolling noise from freight cars is at least 8 dB(A) higher, at the same speed, compared to modern passenger coaches. The cause of this difference is found in the roughening of the wheel tread because of the application of cast iron tread brakes (8). This cast iron tread brake system is the standard brake system for freight wagons in Europe except for the UK where disc braked freight cars form the majority. In the EU-sponsored project Eurosabot (8) new brake blocks are in development that can be easily exchanged with the currently used cast iron blocks, to avoid high implementation costs, and which do not or at least at a lower extent roughen the wheel tread.

Other ways to lower the noise emission of freight cars can be divided into the following three categories:

- avoiding of the excitation which means in fact lowering the wheel roughness. This can be achieved by applying other brake blocks or by applying brake systems which do not interact with the wheel tread like disc brakes, drum brakes or even rail brakes.
- lowering the response of the excited wheel by increasing the internal damping or by applying a geometrically optimised wheel shape.
- applying skirts to avoid the emitted noise to travel directly to the surroundings.

All these items have been studied in the Dutch national project Silent Train Traffic (9). The scope of this project is to build a prototype of a low noise freight wagon and of a low noise track. The combination of these two should give a reduction of at least 10 dB(A) at a speed of 100 km/h. The prototypes will be tested at the end of 1998.

3.5 Operational conditions

With dedicated infrastructure for freight transport the need of a high maximum speed is no longer present. This need was based upon the minimal obstruction of passenger trains which are operated at the conventional lines at speeds between 120 and 160 km/h. The maximum speed INNOCAR is designed for is set to 120 km/h.

Except for the raising of the maximum train speed there are other ways to lower the net transport time like:

- using interoperable locomotives which eliminate the need for a time consuming exchange of the locomotives at the national borders;
- having a dedicated infrastructure;
- giving the same priority to freight traffic as to passenger traffic; this prevents the freight train stopping frequently.
- a good co-operation between the European railways;
- having one train operator who is in charge of the whole journey.

4 Benefits

The benefits of the INNOCAR concept can only be given in a qualitative way. For a cost - benefit analysis a more thorough study is needed which concerns the complete transport chain. This means analysis of all the costs and parties involved. Such a quantitative analysis was felt to be outside the scope of the performed study.

4.1 Foreseen benefits

In qualitative terms the following benefits are expected from operating the INNOCAR on the European Freight Freeways:

- less personnel costs because of the automation of the technical inspections, the coupling and uncoupling process and the setting of the container fixation points and because of the reduced logistic considerations during loading.
- shorter presence at the freight terminal which means that the capacity of these terminals will be increased by operating the INNOCAR;
- an electronically controlled brake system insures a shorter braking distance which allows closer train succession which in fact increases the track capacity;
- shortening of the gross transport time by shortening the loading, unloading and shunting processes;
- designed for the future in regard to the noise emission limits which will come into force within a period of ten year (6);
- better ways to inform the client about the exact time of arrival or the reason for a delay;
- less maintenance costs by switching to condition dependent maintenance based on a diagnosis system;
- energy saving by lowering the tare weight of the wagons.

5 Concluding remarks

From the study performed it is clear that there is a certain need for innovation in rail freight transport. Dedicated rolling stock and infrastructure, interoperability of locomotives and a better co-operation between railways are all ingredients which help the rail modality to profit from the growth of total freight volume to be transported throughout Europe.

In order to have new freight cars in operation which are more efficient and less noisy than the existing wagons there is a role to play for all the parties involved:
- the railway industry has to develop rolling stock like the INNOCAR and car systems like track friendly bogies, non tread brakes for freight cars, lighter vehicles, automatic couplers etc;
- the railways have to invest in new rolling stock to satisfy their clients and to lower costs;
- the national governments should consider subsidising investments in rolling stock which makes them less noisy.

Finally NSM Engineering is looking for a co-maker and an owner to put this INNOCAR concept on the rails.

Artist impression INNOCAR

6 References

1 I. Korpanec and K. Geveke, `European Rail Freight Research', In: *Proceedings World Congress on Railway Research 1996*, p. 13 - 20.

2 P. Fabel, `Increasing the flexibility of freight traffic; using modular train units as an example', In: *Proceedings World Congress on Railway Research 1996*, p. 281 - 289.

3 B. Jahnke, `Automation in rail bound freight traffic' (in German), In: *Schienen der Welt*, 1996, p. 49 - 56.

4 G.J. Bazuin, *Innovations in the rail bound freight transport; an inventory of current developments* (in Dutch), NS Materieel Engineering, Oo/96/04/09, Utrecht, 1996.

5 G.J. Bazuin, *INNOCAR; an innovative carrier for the transport of containers* (in Dutch), NS Materieel Engineering, Oo/96/04/12, Utrecht, 1997.

6 G.J. Bazuin and P.H. de Vos, *Railway noise regulation; an inventory of existing and proposed principles*, NS Materieel Engineering and NS Technical Research, RD/GJB/980129/01, Utrecht, 1998.

7 Work initiated by UIC Sub Committee 25/B *Freight wagen technology*

8 Bazuin (ed.), *State of the Art; Final report*, Eurosabot-consortium, 1N6G30T1.DA, Utrecht, 1996.

9 Silent Freight Traffic, Plan of action ICES-STV, NS Corporate Development, CD/95/V.6.10/038, Utrecht, 1995.

C552/036/98

Delivering just-in-time performance: applying just-in-time logic to rail and intermodal operations

P FOYER BSc, DAE, MIOM, CEng, FIEE
School of Engineering, Coventry University, UK
P N MORTIMER
Track Train Developments Limited, Bognor Regis, UK

SYNOPSIS

Manufacturing, retail and service businesses see time compression as a competitive issue. They are using it as a key means of being more responsive to customer demand at the same time as reducing costs significantly. They would be unlikely to forego these gains in any future change of strategy or transport mode.

The times, quality and costs of logistics (transport, intermodal handling, sorting and storage) are a crucial factor in overall response times, inventory levels and avoidance of losses through spoilage, theft and redundant stock.

Road transport has been attractive for the last 30 years because of its flexibility of transit, low attributable overhead cost and scheduling flexibility. However, it is now beginning to look unattractive because of deteriorating urban and inter-urban transit speeds and increasing unreliability due to increasing traffic congestion. The wider geography of sourcing, by both retailers and manufacturing, has also made average speed and overall journey times a critical issue.

Rail and intermodal have a real opportunity if their operations can become genuinely responsive and just-in-time in their own practices:

- flows in real time and in natural customer demand sequences
- overall short door-to-door times

- unfailing reliability
- quality: dependability, traceability and security

This paper examines the case for smaller, modular trains and 'soft' timetabling and their implications for traffic management, communications network management and vehicle control systems. TruckTrain Developments Ltd and Coventry University have been engaged since late 1997 in project and research work concerned with both the vehicles and their impact on the infrastructure.

1   INTRODUCTION

Supermarkets and major manufacturing industry have made step change gains by the adoption of the Just-in-Time paradigm. They are now able to offer radically wider ranges of products, on demand, to an even more discerning customer base. At the same time, their use of investment, quality costs and overhead costs have all improved radically.

Their main exposure in trying to improve this situation is their dependence upon road transport. This itself is under pressure through increasing congestion, restrictive legislation and diminishing opportunities for productivity improvement; both transit times and costs are set, if anything, to increase over the next few years.

Despite this, rail and intermodal transits have so far failed to penetrate these industries. The reasons are essentially their apparent incompatibility with Just-in-Time. Yet the gains from J.I.T. for supermarkets and manufacturers have been so great that they are, rightly, unwilling to forego them.

This paper argues that if rail and intermodal shippers and network operators applied Just-in-Time philosophies to their own operations, they could compete on more than equal terms.

2   JUST IN TIME PERFORMANCE

Just-in-Time operations are characterised by six key performance improvements:

- goods available on demand in the exact quantities and exactly when needed

- low wastage of materials and time through avoidance of over-production, obsolescence and error

- low capital investment through low inventories and flexible facilities

- low management cost through simple systems, few changes of direction and high quality

- responsiveness to change by both customers and suppliers

- 'quality is free'

2.1 Available on demand

This means exactly what is says: goods should be available when the (end) customer is ready to use or buy them

The customer decision can be either before of after the point of production and/or delivery.

The implication is that manufacturing and delivery processes should take place as near in time as possible to the point of the customer making his decision. Too early involves risks of production and delivery of unwanted goods. Too late means the customer is dissatisfied or even loses tangibly. Closing the gap between the demand-side decision and supply-side action is crucial.

2.2 Low waste

Goods, time and resources are wasted whenever the 'best' timing match between demand-side decision and supply-side action is violated. These often small errors for small amounts of material tend to carry cost penalties several orders of magnitude larger than their apparent size. For instance, one component delivery failure can cost as much to correct as organising a whole day's production, sales or deliveries. Chain reactions can spread to waste people's time and other material costs and miss overall value-adding opportunities.

2.3 Low capital requirements

The use of Just-in-Time practices throughout the chain means that the assets needed to support the operation are massively reduced: working capital, storage, production and sales space, control equipment, communications. Getting control of lead-times can release not only large amounts of capital but also expensive space to increase production or sales.

2.4 Low overhead costs

Supermarkets and manufacturers tend to look hard at the theoretical material and labour content of their products. If everything goes well, then this leaves a large Gross Margin to service capital investment and pay for management. Just-in-Time reduces wasted materials and time, as well as excess investment, which are the main controllable elements in the burden costs which dissipate the Gross Margin.

2.5 Quick response to change

All businesses need to respond to change: not only to longer-term underlying change, but also to short-term fluctuations in demand, supply and product ranges.

Systems which are full of work-in-progress (tangible and brain-work) and forward commitments are inherently both slower to respond to change and more likely to incur major redundancy of unwanted materials and work.

Just-in-Time systems are (or should be) essentially 'empty', with minimal bureaucratic baggage; as a result they are able to respond quickly with low fall-out of frustrated effort.

The Set-up Cost (cost of changing type) becomes a major target in Just-in-Time programmes of improvement, so that the sequence of work can be demand-led rather than constrained by the cost of changing over.

The benefits of Just-in-Time can be characterised by 'Economies of Scope'.

3   THE CHALLENGES OF JUST-IN-TIME FOR RAIL AND INTERMODAL OPERATIONS

Rail and intermodal operations have have grown up with the 19th Century paradigm of 'Economies of Scale'. The characteristics of this were to stereotype and herd apparently similar things into maximum-sized groups and so spread set-up costs. The penalties for this apparent benefit were poor service, administrative complexity and often poor quality.

In this context, containers,with at least as much capacity as large road vehicles, can be a severe constraint on flexibility; a train has a massive capacity to fill.

To adapt the train/container to the Just-in-Time paradigm, the challenges to be met include:

- cutting the maximum (as well as minimum) time between production and customer decision

- changing production and delivery between types and sources to meet short term change

- keeping investment in all assets (production, transportation, storage) to a viable level

- keeping management costs minimal despite the increased complexity of intermodal operations

- adapting transport rapidly to changes in demand and supply

- delivery quality so good that quality-caused failures are no longer significant causes of disruption

3.1 Time between customer decision and production

The time between the customer decision and production is typically made up of:

- actual production time

- actual journey time

- time to accumulate and finish the production batch size (essentially of the same type of goods)

- time to accumulate and distribute the delivery lot size (essentially between the same destinations)

- departure frequency

- load, unload and intermodal transfer times

- lead time to book a path or journey

All of these (except perhaps actual production time) need to be radically reduced to make rail or intermodal competitive for the majority of inland applications. Journey times for freight trains need to become at least as fast and predictable as for passenger traffic. Departure frequency needs at least to match the rescheduling frequency of modern supermarket and vehicle build operations (typically several times per 24-hour period). Load, unload and intermodal transfer times need to be improved until they become an insignificant part of the source-destination journey time (as opposed to the 1⁄2-day needed to load or unload a 2000-ton container train at present). It must be made possible to book or re-book spaces on trains and/or train paths at short notice (minutes or a couple of hours at most). The analogies to airline seat reservations and flight slots are clear but have yet to be exploited.

3.2 Changing production and delivery sources and types

Supermarkets and assemblers often service similar items from more than one source, in order either to provide sufficient capacity, secure reliability of supply or offer different options to their customers. This means that an apparently steady inward flow may fluctuate rapidly between different sources.

3.3 Keeping investment low

Rail assets, as well as storage (particularly temperature-controlled) are capital-intensive. The real cost challenge is simultaneously to keep inventories low and to utilise transport equipment intensively. This latter is currently true of passenger rail vehicles but not of rail freight equipment.

3.4 Keeping management costs down

Rail and multi-modal operations are inherently more complex than traditional road haulage. Equally, road hauliers are having to increase their management burden costs to offer acceptable reliability, traceability and security for their customers.

The challenge to rail and inter-modal operators will be to obtain genuine on-demand operation, reliability and security at the same time as bringing management burden costs down radically.

3.5 Quality

The quality of distribution is measurable by two key criteria:

- reliability, flexibility and predictability of service

- condition and security of goods delivered

Rail arguably has a better infrastructure than road, but often fails to deliver on either or both measures above.

4   POTENTIAL SOLUTIONS

It is likely that the rail and inter-modal industries are fully capable of delivering solutions to all of the challenges, given the necessary business will and the operational skills development.

4.1 Lead times

This measure is a fundamental indicator of capability and change. There are two kinds of potential solution approaches:

- to increase the frequency and speed of multi-consignment trains and allow quick and easy access at pre-planned stopping points for loading and unloading (like a passenger train)

- to decrease train size and increase the departure frequency to the point where one train would be fully loaded and utilised for one (customer's) traffic flow

Both are possible, but require other challenges to be addressed first.

Multi-consignment trains will require action to open up Booking Systems, speed Intermodal Handling and solve some serious Security issues.

Small trains (such as TruckTrain) will present pathing problems in some areas, require seamless and interactive Path Booking Systems and involve reliable security and condition-monitoring systems.

4.2 Frequently-changing production and delivery sources and types

This requirement obviously favours short trains, but may place severe problems on Path

Booking systems and, in more extreme cases, challenge Driver Route Knowledge. Where sources and destinations are not rail-connected, the ability of Path Booking Systems to organise seamlessly and simultaneously the necessary road and rail transits will be vital.

4.3 Keeping Investment Low

This requirement implies a need to keep both materials and vehicles moving 24 hours per day, often 7 days per week. In this way, warehouse space can minimised, as can numbers of rail and road vehicles and intermodal transfer facilities.

A total change of paradigm in the management of rail freight vehicles will be needed, both to achieve the necessary intensive scheduling and to make their maintenance reliable enough to allow them work continuously for several weeks between servicing.

A well-planned sharing arrangement for vehicle control between rail traffic management, road operations and users' operations management will be needed to achieve the requirements for Path Booking and Security, Condition Monitoring and Traceability at acceptable cost. Ideally the maximum functionality should be on the vehicle and with the Shipper/Vehicle Operator to allow maximum flexibility and clarity of responsibility.

4.4 Keeping management costs down

Low management costs depend on 'right-first-time'. This means both avoiding calls for redundant transits and achieving absolute reliability of transit.

One implication will be that the Path Booking System needs to be able to operate at very short notice (and totally visibly and in real time).

The other key implication is that improvements in both the reliability of vehicles, infrastructure and their ability to cope with and recover from disruption (again with real-time information) will be critical. The cost of disruption is a major Quality (waste) cost to both the Network Operator, the Vehicle Operator and the Customer.

4.5 Quality

Apart from making unnecessary production and/or transits, the most important business improvement target must be quality.

Some of the issues are cultural; industry has been solving these problems by combinations of organisational change and improved managerial behaviour. The formation of teams with responsibility and scope for delivering a whole product or service to the customer (the 'Cell' concept) has been the most fundamental change in making ordinary work people and their managers accountable for tangible value and costs.

Technically, operators of vehicles, terminals and systems can take simple steps to improve transit quality:

- install condition monitoring systems on vehicles (for traceability, manifest integrity, load condition, staff and load security and functionality)

- set up automatic GSM/GPS phone links to raise support in emergencies and install 'Black Boxes' to record events including management actions and vital signs reports

- protect areas where vehicles are 'open' or unmanned with security enclosures and electronic security

- screen employees, use secure identification for authority to move or unload vehicles and monitor all communications to or from vehicles

- provide on-board facilities to allow the use of alternative routes (eg on-board location and navigation equipment, on-board visual simulator displays)

Few transit companies or network operators have 'Cost of Quality' monitoring and management systems, but the true costs of even small incidents can be a significant part of total overhead burden costs (ie suck out significant proportions of the Gross Margins)

5   CONCLUSIONS

Rail and intermodal transport have the potential, with their controlled environment and potential for high average speeds, to deliver real advantages to many types of manufacturing and retail businesses.

The key targets for improvement to realise this potential are a direct correlation from industrial Just-in-Time practices:

- **to allow greatly increased departure frequencies**

- **to improve loading, intermodal transfer and unloading times and techniques radically**

- **to provide an on-demand booking system for either unit loads or paths for short trains**

- **set up reliable condition monitoring and communications on all vehicles and at all staging points to improve transit quality and security**

- **organise staff into teams charged with near-complete customer transits to improve 'ownership' of customers' interests**

- **invest and improve processes to achieve genuinely excellent quality and security, as investments which will tangibly and quickly show good returns.**

6   REFERENCE

'The Machine that Changed the World'; Womack, James P, Jones, Daniel T, Roos, Daniel; Rawson Associates/ Maxwell Macmillan 1990; ISBN 0-89256350 8

C552/046/98

Developments of the TF25 'track-friendly' bogie

A B HARDING BSc, CEng, MIMechE
Powell Duffryn Rail Projects, Cardiff, UK

Synopsis

In 1997 Powell Duffryn Rail Projects set out to re-engineer the Low Track Force bogie transforming it from a highly individual, high cost, high tech freight bogie into a simplified, mid cost bogie which still delivered outstanding performance.

This has been largely achieved with the development of the Track Friendly TF25 bogie which currently encompasses two brake variations, push brake with off bogie actuation and integral push brake. A third variation for disc brake is under development. The process of UIC homologation has also commenced. The next stage will be to review the design, simplify again and further reduce the cost.

This paper sets out this process of transformation from high tech, high cost to low tech, low cost whilst retaining track friendly properties.

Introduction

Whilst the LTF25 bogie has established beyond doubt the benefit of Low Track Force technology to the railway, it became increasingly clear that, due to its high tech image and high first cost, the bogie was never going to be universally adopted in its existing format. The task ahead was to redesign the bogie to retain its low track force benefits, including the Track Access Discount from Railtrack whilst significantly simplifying and reducing the cost of the bogie.

1 THEORY OF THE SUSPENSION

In the early 1980's it was noticed by the B R Civil Engineer that the Western Region freight up-line was deteriorating much faster than the down-line. This was quickly attributed to the laden stone trains from the Mendip quarries carrying aggregate to the south-east and returning empty.

As a result of this the engineers at B R Research were asked to develop a freight bogie which, whilst being more friendly to the track at 100 km/hr, would enable 25 tonnes axleload to run at 120 km/hr without causing additional damage to the track, hence the concept of Low Track Force bogies were created.

It was well established that, at the wheel to rail interface, the forces could be broken down into static (P_0) forces and dynamic (P_2) forces. For a given axleload it is only possible to modify the P_2 forces. On the Low Track Force (LTF) bogie this reduction in P_2 force is achieved by reducing the unsprung mass of the wheelset (by reducing the wheel diameter and using inside journals) and eliminating friction damping at the primary suspension, thus peak forces are significantly reduced. Although the wheel diameter is reduced, the tread contact stresses are maintained at acceptable levels by maintaining a wheel contact patch of sufficient size by the selection of a suitable wheel profile.

At this point, the wheelset is unguided by the suspension therefore it is necessary to hold the wheelset square to the bogie frame by traction rods. As the wheelset is much more compact in plan the inertia forces are similarly reduced which enables relatively soft end mountings to be used in the primary traction rods. The benefit of this is that in normal running the wheelset is very stable but, due to the reduced stiffness of the mountings it has very good steering characteristics.

The relatively stiff vertical primary suspension only accommodates track twist and other track irregularities, therefore the main suspension is provided by rubber 'hourglass' springs located at the secondary position. These springs permit vertical, lateral and yaw movement of the bogie under the vehicle. Damping is provided also at the secondary position by inclined, load sensitive hydraulic dampers giving both vertical and lateral control. The bogie does not have a conventional centre pivot and sidebearers, but rotates by the longitudinal displacement of the rubber secondary springs which provide a restoring moment on exiting curves. A single secondary traction rod controls traction and buffing loads.

Figure 1:   LTF25 Bogie

2 BACKGROUND OF THE LTF BOGIE

As a result of this new approach, **trials** have shown that the **dynamic vertical track forces** have been **reduced by 30%** and **lateral track forces by 60%.** Measurements of acceleration on a masonry bridge showed those LTF25 bogies **at 25.5 tonnes** axleload produced about **half the acceleration** of Y25 bogie at **22.5 tonnes** axleload. The difference between the LTF25 and the Y25 is more significant on smoother track and is particularly marked on good quality intercity track, and it was noted that the **dynamic forces** from **LTF25** bogies **are no higher at 120 km/hr** than at 100 km/hr whilst the forces from the **Y25 bogies increase** by about **10%.**

Due to the many degrees of freedom of the suspension the bogie will curve freely down to a radius of 400m. This is borne out by service experience with one fleet of 102 tonne GLW stone hopper wagons that have covered, on average, 350,000 kms before first re-profiling.

In addition to the reduction in track forces, railborne noise has been significantly lowered with reductions of between 15-20 dB(A) being recorded. This has been achieved not only by the reduction in vertical forces but also better steering eliminating flange squeal, and of course disc braking rather than conventional cast iron tread brakes.

This all seemed very promising, and whilst the first cost was high, early sales indicated that the more sophisticated buyer was prepared to look at the whole life cost rather than at the very narrow view of first cost only. As is often the case, particularly in the UK, the bogie was developed in production and a series of component failures dogged the project. Whilst these were all resolved, and substantially increased our knowledge of the product, it held back the widespread introduction of the bogie. European railways, whilst interested in the bogie, were concerned about the operational problems created by inside journals and the introduction of Eastern European built bogies reduced the base price for freight bogies still further.

In this climate it became increasingly clear that LTF bogies, at least in their purest form, were never going to gain widespread acceptance. However we had built up a large database of information on low track force technology and we felt we could re-engineer the bogie to become both technically and commercially acceptable to the majority of European railways.

3 DECISIONS IN A CHANGING MARKET

Having recognised the need for change we commissioned a European market survey into market drivers for freight bogies. Not surprisingly it showed that the overwhelming factors were:-

- **Low first cost**
- **High reliability**
- Heavy payload
- Long product life
- Ease of maintenance

The first two factors outweighed all the others, and all the good things we felt we had to offer with LTF were rated very lowly. However we were equally aware that there were other pressures on the European railways such as the increasing pressure to lift axleloads to 25 tonnes and the onward march of the anti-noise lobby who were pushing through legislation in various European countries.

At this point we decided to set ourselves a list of targets, both technical and economic that we now felt were essential if we were to succeed with a new bogie.

4 INITIAL TARGETS

Although never stated explicitly it was always our intention that any replacement for the LTF bogie would largely retain its track friendly properties at 25 tonnes axleload and potentially upto speeds of 120 km/hr. The new targets we set ourselves were:-

- **50% Cost reduction**
- **10% Weight reduction**
- **75 dB(A) Noise level**
- **60% Maintenance costs**
- **99% Reliability**
- **Patents**
- **UIC Compatibility (If possible)**

The first two items relate to reductions against the existing LTF bogie, whilst the reduction in maintenance is against the industry standard bogie, the Y25. All the other items are self-explanatory.

5 REVIEW OF DESIGN OPTIONS

At this time we still felt it was necessary to retain the inside journals to achieve low track force even though we were aware that it was a drawback to receiving approval. We produced a number of different proposals, some of which were alarmingly radical, and were rejected at an early stage. We sought outside opinions on one or two of the designs, mainly because we still felt uncomfortable about the concepts involved. It soon became apparent that we would have to 'sacrifice some sacred cows' if we were to really make the breakthrough we needed.

This decision freed us to be really radical (from our perspective) and within a very short time we had abandoned both inside journals (together on-board hot box detection) and hourglass springs. Whilst this simple decision immediately made the bogie more acceptable and cheaper, it did not however provide us with a ready made design.

The real breakthrough came when we decided to reverse the suspension! The **original LTF** bogie has a **stiff, undamped primary suspension** with a **soft, hydraulically damped secondary suspension**, we decided to reverse this by creating a **new bogie** with a **soft, hydraulically damped primary suspension** and a **stiff, undamped secondary suspension**.

Running this new arrangement through the Vampire dynamics modelling package showed us immediately that we had a viable suspension that was definitely track friendly but significantly simpler than its predecessor. The concept of the Track Friendly TF25 bogie had been created.

Figure 2:   TF25 Bogie

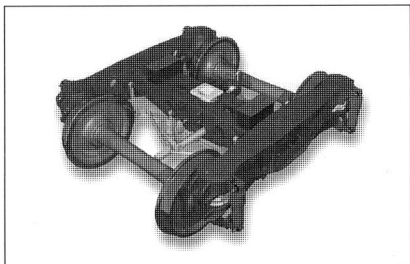

6 DETAIL DESIGN

We now had little more than a concept with promise, the make or break would come with the detail design. It was decided from the outset that we would avoid any unnecessary bogie mounted equipment such as handbrakes that would make the bogie both more expensive and more complicated. It was also decided at this point that each detail change to the bogie, where appropriate, would be fed into the Vampire dynamic model to ensure that even relatively small changes were not going to have an effect on ride or track forces. Similarly structural changes were fed back into the FEA model to make sure there were no nasty surprises at the end of the design process.

We were now at a point where we could make some fundamental decisions about the detail design of the bogie. Firstly for maximum manufacturing flexibility we needed both cast steel and fabricated steel bogie frame versions, and those frames and their component parts must be fully interchangeable. We decided we would design for minimum components and de-skill the assembly process by removing any opportunity for misinterpretation of instructions on the shopfloor. For example we have selected the use of Huck bolts for several fixings rather than conventional torqued bolts as the latter can be fitted incorrectly or interfered with in service.

A search for patentable features revealed several promising possibilities and we have currently four separate applications on file.

7 WHEELSETS

We could now concentrate on the actual detail of the bogie and naturally enough the first item we reviewed was the wheelset. To minimise unsprung mass it had already been ascertained that, as there was a two stage suspension, an 840 mm diameter wheel could be used rather than the conventional 952 mm diameter normally used for 25 tonnes axleload. The principle of smaller wheels had already been well established by the LTF25 bogie where we had successfully used the even smaller 813 mm diameter wheel. The 840 mm wheel is also a UIC norm, albeit for a lower axleload.

Having previously decided to revert to outside journals the axle design was conventional and we elected to fit Timken 150 mm diameter cartridge bearings fitted with the latest HDL seals. Although, at this stage, we are using forged and rolled steel wheels provision has been made to fit the wider American cast steel wheels. Finally, for the UK, we have chosen the widely used P6 profile rather than the P8 passenger profile used on the LTF25 bogie. This provides us with good stability upto 120 km/hr and maintains an adequate contact patch to keep contact stresses at an acceptable level. P6 is widely used as a freight profile and will assist ready access to all shops where re-profiling is carried out.

8 PRIMARY SUSPENSION

During the theoretical phase of the design we had established the vertical and lateral characteristics of the primary suspension, it was now necessary to provide the hardware to give the required performance. We selected a radial arm arrangement, which minimised the number of components and by careful positioning of the pivot point ensured that the coil spring assembly underwent the minimum longitudinal shift due to the rotary action of the arm. The radial bush was selected to give the right combination of lateral and longitudinal movement when taken in parallel with the main coil springs. This assists the lateral ride and gives an element of controlled steering of the wheelset assembly. The radial arm bush is fastened to the sideframe semi-permanently by Huck bolts.

Hydraulic dampers mounted between the sideframe and radial arm provide vertical damping. The hydraulic dampers are fitted with rubber end bushes that have shafts with a trapezoidal section for fixing to the end mountings. This provides a secure fastening which can be easily demounted on the lineside if necessary. The whole assembly is secured by a lift stop which doubles as a lateral bump stop.

The radial arm forms an open saddle at the journal into which the bearing fits enabling the wheelset to be removed without the need to disassemble the suspension. The wheelset is secured by a simple hinged strap, which is fastened by a single bolt.

Figure 3: View of Primary Suspension

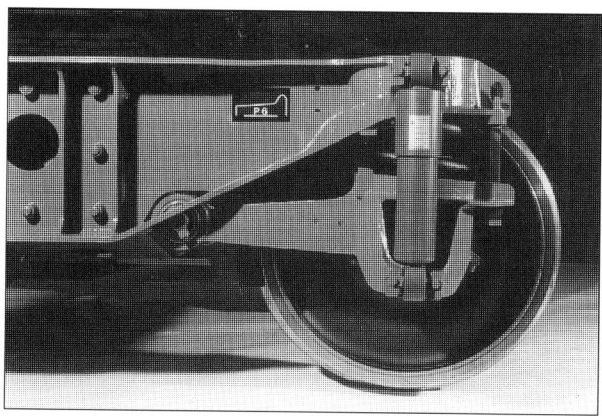

9 BOGIE FRAME

The bogie frame is a rigid 'H' frame made up of two sideframes and a bolster which are fastened together by Huck bolts. The bogie frame is designed to provide the option of either cast steel or fabricated steel to suit the preference of the customer. These components are fully interchangeable and a frame theoretically could be made up from a cast bolster and fabricated sideframes or some other combination. In the event of a damaged frame those elements damaged can be removed and easily replaced by new items as all interfaces are machined to guarantee interchangeability.

The frame has been designed to load cases as specified in UIC document RP17 and the stresses verified by finite element analysis and fatigue analysis where appropriate. Both of the frame options, cast and fabricated, will be fatigue tested to RP17 to meet UIC structural requirements.

Figure 4:   Bogie Frame Joint.

10 SECONDARY SUSPENSION

The secondary suspension consists of two secondary springs and a traction centre to control the lateral, longitudinal and yaw motions of the bogie.

Each secondary spring, whilst vertically stiff, needs to accommodate 35 mm lateral movement to give a good lateral ride and to shear sufficiently longitudinally to permit bogie rotation on the minimum curve of 60 m. There is minimal vertical deflection. Metalastik have developed a patented design exclusively for the TF25 bogie which consists of rubber/steel chevron sandwich with a very shallow included angle and the requisite characteristics are obtained by coring out the rubber. The whole unit is attached to the bogie by two dowels. The secondary spring occupies the space normally taken up by the sidebearer but is rigidly fixed to the wagon by dowels. There is no slipping movement between the wagon and the spring and as such no friction liners to wear out.

It was apparent to us that one of the drawbacks of the earlier LTF bogie was its attachment of the secondary traction rod to a large bracket fixed to the wagon. We were therefore determined that the attachment of the wagon to the TF25 bogie should be as easy as a conventional bogie. The traction centre which is in two parts replaces the conventional centre pivot and the lower section consists of a steel centre to which is attached a horizontal lateral damper and the longitudinal secondary traction rod.

The lateral damper, naturally enough, controls the lateral motion of wagon body sitting on the secondary springs. Currently this is a single rate damper but for large tare to gross ratios it is planned to introduce a variable rate damper activated by the load weigh valve.

The traction rod is attached to the bogie at one end via an external bracket which doubles as a brake lever fixing and to the traction centre at the other end. Each end is fitted with a resilient bush and the whole system is designed to withstand 5g.

For ease of assembly the traction centre has an easily detachable upper section which is bolted to the underside of the wagon bolster. The upper section fits into a receiver on the lower section in the same way as a normal centre pivot. In this way demounting the body from the bogie is identical to a conventional bogie with a centre pivot and sidebearers.

Figure 5:   Secondary Suspension

11 BRAKES

To date we have identified three different brakes which we believe we need to develop.
These are:

- **Tread brake, wagon mounted actuation**
- **Tread brake, bogie mounted actuation**
- **Disc brake, bogie mounted actuation**

It would also be possible to fit a standard cast iron block clasp brake arrangement but it
was felt that as this would be heavy, noisy and potentially expensive. We would only produce
it if specifically requested by a customer.

Due to it's simplicity it was decided to fit the tread brake system to the prototype bogies.
The brake consists of a four block push brake arrangement with brake actuation from a wagon
mounted brake cylinder and slack adjuster.

12 DESIGN REVIEW

In order to facilitate an early test date it was decided to freeze the design at a certain point in
time. This meant that in some instances there was still some refinement of the detail design
outstanding. This we felt was acceptable as the manufacturing process, the scrutiny and an
FMEA we commissioned from Manchester Metropolitan University would throw a series of
areas for further consideration. These points have now been dealt with during the design
review process.

13 REVIEW OF SUPPLIER OPTIONS

It was obvious to us from an early stage of the project that, not only would we have to be innovative in our approach to the design, but also need to review our sourcing options. The major competition for freight bogies in Europe is either from low cost sources in Eastern Europe or more recently the USA.

Costings we had already obtained showed that it was unlikely that the new bogie would be competitive in the European market if it was manufactured in the UK or contained a large proportion of British parts. It was an unpalatable fact but one that could not be escaped. Fortunately we had already built up considerable expertise in sourcing from Eastern Europe.

We started the supplier search at an early stage and used our existing contacts to provide us with lists of potential suppliers who were already supplying to the railway industry in Europe. This process threw up some new names to us and trial quotes plus a QA visit soon sorted out the companies that merited further investigation.

Our initial search provided us with a good selection of potential suppliers therefore, so as not to give ourselves too big a task, we decided to divide them into two categories, those we wished to develop immediately and those that we would need a higher degree of confidence before we brought them on stream.

In the latter case this did not necessarily reflect on the ability of the supplier, although this clearly had an input, but was more to do with the classification and degree of specialisation of the component. The failure of a bearing or a damper, for example, early in the introduction of the TF25 bogie could have a major impact on the viability of the whole project. Therefore in the medium term it was deemed to be prudent to stay with tried and tested suppliers for those components.

We have now developed a good portfolio of suppliers across Europe including a wheelset manufacturer, malleable iron and steel foundries, steel spring manufacturer, fabricators and a source of brake systems.

All of these suppliers operate to the internationally recognised quality standard ISO 9000 and above and most have approvals from one or more national railway authority. None of these companies are registered as an approved supplier until they meet the minimum standard set by our own quality department.

Working with these partners we are constantly striving to drive costs down by jointly value engineering the individual components whilst ensuring we maintain or improve product quality.

Figure 6:   TF25 Bogie

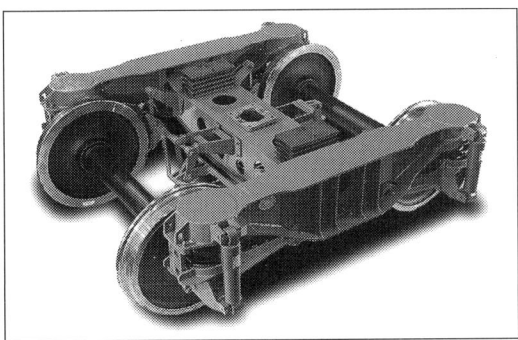

14 EVALUATION AND TESTING

At the time of going to press the first two prototype bogies are complete and fitted under a 102 tonne bogie box wagon which will go to Derby for ride and brake trials during October and November. A second pair of bogies are being produced to be fitted to a tank wagon for trials in the UK, whilst a third pair of bogies are being produced in the Czech Republic to be fitted under a sliding wall van for trials in Europe.

AEA Technology have undertaken a study of comparative track damage for a variety different freight bogies commonly used on Railtracks network. Not surprisingly the full LTF bogie shows the least damage, nearly 30% better than the worst primary suspension bogie. The TF25 bogie follows as a close second to the LTF bogie being only some 3% less track friendly. As a result Railtrack will shortly confirm that the 10 % Track Access Discount they currently offer for use of the LTF bogie will be extended to cover the TF25 bogie.

At the beginning of this exercise we set ourselves a number of objectives, namely:

50% Cost reduction

Achieved: As further evaluation is undertaken we expect maintain and/or improve this situation.

10% Weight reduction

Achieved: Although the first two bogies exceeded this target we know where the excess weight is located and this will be corrected on the production bogies without detriment to strength.

75 dB(A) Noise level

To be established: As the bogie has a soft two stage suspension incorporating rubber elements it should have similar noise characteristics to the earlier LTF bogie which is known to be very quiet.

60% Maintenance costs

To be confirmed: Analysis of maintenance requirements show that apart from the normal consumables, brake blocks, wheelwear etc the bogies will only require attention during the annual VIBT inspection. Due to the limited self steering ability wheelwear will generally better than a Y25 but not as good as the full LTF bogie.

99% Reliability

To be confirmed: An FMEA carried out on the bogie as identified those components that carry a high risk of failure and additional work has been done to reduce those risks. More fundamentally by reducing the total number of components in the bogie it is easier to identify high risk items and increase the design effort to maximise reliability.

Patents

Achieved: We have identified four separate patentable features and European patent applications have been lodged. Additionally one supplier has designed a novel component which is incorporated into the bogie and we have applied for a joint patent.

UIC Compatibility

Not achieved: To produce a bogie that is track friendly, and with 25 tonnes axleload at 120 km/hr, it has not, at this time, been possible to respect all of the UIC norms that would make the bogie UIC compatible. However it is possible to fit the TF25 bogie under existing wagons with minimal rework

Although not all of the initial targets have been wholly achieved we consider that we have got the fundamentals right and the project is a success. We have already identified those areas where more work is required and we see product development/improvement as an ongoing task.

It has taken 15 months of fairly intense activity to transform an identified need for our business into a fully working marketable product which looks set to have a long future in the UK and European railfreight industry.

C552/028/98

Improving ride quality for the transport of automobiles

K ROWND and **C URBAN**
Transport Technology Center Inc., Pueblo, USA

SYNOPSIS

Partnerships between the railroad and automotive industry have been established to identify ride quality performance objectives for transporting finished automobiles by rail. Tests conducted by Transportation Technology Center, Inc. (TTCI), a subsidiary of the Association of American Railroads (AAR), indicate that these ride quality improvements can be met by using advanced suspension designs and conventional multi-level autorack rail cars. Testing with modified conventional suspensions has demonstrated that tuning existing designs for the requirements of autorack service results in improved performance.

Ride quality tests were performed at the Federal Railroad Administration's Transportation Technology Center (TTC) and in nearly 6,000 miles of railroad service for each suspension system. Tests were performed in partnership with TTX Company as part of the AAR Advanced Freight Car Truck (Bogie) Program. The goal of the Advanced Freight Car Truck Program is to promote innovative suspensions for freight cars based on commodity-specific requirements. The Quality and Maintenance of Equipment working group, comprised of automobile and railroad members, has supplied additional support.

New criteria for ride quality performance are contained in Recommended Practice (RP) 803-96, *"Ride Quality Performance Requirements for Motor Vehicle Shipments."* The automobile and railroad industries have considerable investment in multi-level autorack railcars and associated facilities. If ride quality requirements can be met with the existing fleet, multi-level railcars will remain the dominant mode of shipment for reasons of cost and efficiency.

1. INTRODUCTION

Three-piece bogie designs have been used by the North American railroad industry for many years. However, as demands for dependable, reliable, and efficient transportation service increase, innovative suspensions are needed. This is especially true for the transportation of finished automobiles. Although damage to automobiles transported by rail has reduced over the past several years, further improvement is being demanded by the automotive industry. Vertical and lateral ride quality improvements can be achieved through advanced freight car suspensions.

In 1995, the Association of American Railroads (AAR) established a multi-year; industry-funded program called Advanced Freight Car Truck (Bogie) to promote improvements to the general fleet of freight car suspensions. The focus of the Advanced Freight Car Truck Program in 1996 and 1997 was improved suspensions for bi-level autorack cars.

The introduction of enhanced ride quality for transportation of finished automobiles is a customer driven issue. Nearly 70 percent of finished automobiles are transported by rail. The net present value savings for an idealized advanced suspension is approximately equal to that calculated for a 5 percent change in market share. The intent is to assist the automobile manufacturers in achieving the following goals:

- Error-free transportation
- Elimination of rail transportation as an automobile design consideration

Since 1992 TTX Company has spent more than $200 million to improve performance of this service. TTX and AAR have pooled resources to investigate ride quality performance of advanced suspensions under multi-level automobile carrying rail cars.

2. BACKGROUND

Ride quality objectives are expressed in terms of acceptable acceleration performance measured in tests described in a recommended practice for railcars used for automobile transportation (RP 803-96) entitled: *"Ride Quality Performance Requirements for Motor Vehicle Shipments."* For the first time, the railroad and automotive industries have reached agreement on how ride quality data is to be collected and analyzed.

This RP specifies standard test and analysis cases for evaluating ride quality performance. It also describes methods for data collection and analysis. Requirements include the following:
- Specified instrumentation and data collection techniques
- Controlled tests over specially constructed track anomalies to identify design weakness and promote design development
- Railroad service tests on railroad property for designs meeting controlled test criteria

3. ISSUES

3.1 Investment in existing fleet
Effective use of the existing fleet comprised of more than 45,000 multi-level autorack cars and associated facilities is one important issue. Multi-level service is the quickest and most cost-effective method for shipping automobiles by rail.

3.2 Vertical ride performance
In the past, the automotive industry was primarily concerned with the longitudinal and lateral stability of multi-level cars. To achieve improved lateral stability, TTX initiated a program to replace old technology three-piece bogies with premium bogies. The automotive industry has collected railroad service data on cars equipped with the premium bogie. These tests identified the need to improve vertical ride performance.

3.3 Automobile suspensions and restraints
The automotive industry has moved from the traditional chain tie-down system to a wheel chock system for restraining automobiles on a rack car. The chain system effectively locks out the automobile suspension by pulling the body to the railcar deck. This makes response to railcar

movement predictable, but can transmit damaging shock loads into automobiles. The chock system can reduce damage from shock loads; however, it makes rail car response interactive with automobile suspensions. As documented in AAR *Technology Digest* TD97-038, "Improved Ride Quality for Rail Transport of Finished Automobiles," this live load can change the response of the railcar system to track and operating variables. Exhibit 1 shows an automobile restrained by a wheel chock system. Four chocks restrain each automobile.

Exhibit 1. Automobile chock restrain system

4. APPROACH

The approach for the Advanced Freight Car Truck program included the following:
- Documenting expectations with a performance specification
- Soliciting new suspensions with the specification
- Supplying computer models and engineering support in the design process
- Evaluating concepts using a vehicle dynamics computer model
- Design review meetings with AAR, TTX, and each proponent
- Evaluation of prototype performance in controlled tests at TTC
- Railroad service testing of selected candidates
- Evaluation of ride quality performance using criteria described in RP 803-96
- Quarterly updates to a joint auto and rail working group

The performance specification was sent to proponents who previously answered a generic initiative to participate in the program to improve autorack suspensions. The specification addressed safety, ride quality, and economic factors.

TTX supplied a standard autorack for this program. AAR constructed a model of this railcar within the vehicle dynamics simulation program NUCARS. This model was supplied with the solicitation for new suspensions.

Fifteen design concepts were submitted. Proponents (or AAR) performed simulations of the safety and ride quality tests using NUCARS. Several design review meetings were conducted at TTX to discuss practical implementation issues and to review simulation results. Thirteen bi-level prototypes, one baseline, and three premium suspension have been tested at TTC.

A joint auto and rail industry working group, the Quality and Maintenance of Equipment (QME) was used as a forum for disseminating information from this program. QME meetings allow automobile industry representatives to comment on the suitability of the results.

5. RESULTS 1995-1996

Three suspensions were evaluated in controlled tests in 1995: the baseline (old technology), a premium design (Swing Motion) replacing the baseline, and a new design (Krupp TI-7R). The controlled test program demonstrated the following:
- The old technology bogie does not meet lateral or vertical ride quality requirements.
- The premium bogie meets requirements for lateral performance but does not meet expectations for vertical performance.
- The advanced bogie meets both lateral and vertical performance requirements.

5.1 Bogies Tested

The old technology bogie: This suspension served as the baseline for all bogies tested in the Advanced Truck program. This design has a secondary coil spring suspension with friction snubbing. The spring group consisted of seven outer and three inner D-5 coils. The bogie had constant damping and was outfitted with Miner TCC-II-60 constant contact side bearings.

The premium bogie — NACO Swing Motion: A transom acts as a shear plate to increase warp resistance by connecting the bogie sideframes. Special bearing adapters with a rocker seat allows the sideframes to swing laterally. This lateral degree of freedom de-couples the wheel set and bogie motions from the car body lateral motion. The secondary suspension utilized variable friction damping provided by wedges controlled by two No. 49427 coils. Low friction material was applied to the vertical surface of the wedges. The spring nest consisted of six D7 outer coils. The bogie was equipped with Miner TCC-II-60 constant contact side bearings.

The advanced bogie — Krupp TI-7R: The advanced bogie (Exhibit 2) had a leaf spring bolster resting on a spring nest of four D-7 coil springs per side. The D-7s are placed on the transom that sits in the bottom of a modified 70-ton sideframe design. The connections between the transom and the sideframes allow lateral motion. The side bearings were Miner TCC II-35 long travel, set at 4 3/4 inches height under load. The side bearings were attached to the leaf spring bolster. Koni O2A-1374 vertical dampers were attached to the side bearing caps and sideframe.

Exhibit 2. Advanced bogie—Krupp TI-7R

5.2 Controlled Tests at TTC

5.2.1 High-speed stability

The high-speed stability test was conducted over a 5,000-foot smooth tangent track. The criterion for success is a standard deviation of lateral deck acceleration of no more than 0.13 g, as tested at constant speeds from 40 mph to 70 mph. The premium and advanced suspensions met criterion, while the baseline bogie did not. TTX has been replacing the old three-piece bogie design with the premium bogie to improve lateral stability.

5.2.2 Pitch and bounce

The pitch and bounce test is intended to exercise the vertical suspension. A specially constructed track with ten 39-foot vertical bumps, each 0.75 inch amplitude, in phase on each rail is used to excite the rail vehicle. Test speeds are from 40 mph to 70 mph. The criterion for success is that maximum vertical deck acceleration must be no more than ± 0.5 g.

The old technology bogie did not meet performance criterion. Exhibit 3 shows the maximum and minimum autorack accelerations for the premium and advanced bogies at each speed tested. At speeds above 55 mph, the premium bogie did not meet the criterion. The automobile manufacturers have criticized the vertical ride performance of this premium design. The advanced bogie stayed within the performance criterion, indicating improved vertical suspension.

5.2.3 Pitch and bounce automobile response

Exhibit 4 shows the response of a pickup and sedan to harmonic rail car deck accelerations, as tested with the premium bogie. Railroad service will impart discrete as well as harmonic excitation to the railcar. The reaction of the automobile appears to be independent of speed above some level of rack motion. Data for the automobiles, as tested with the advanced and three-piece bogies, shows a similar trend. The sedan had much less response to deck motion than the pickup bogie. Although there is no official criteria for automobile acceleration, this data and data from the twist and roll test, points out the need to consider the automobile suspension when evaluating suspensions for rail cars.

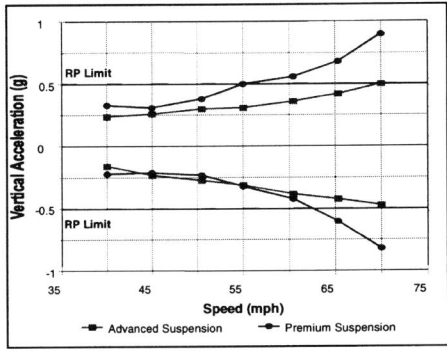

Exhibit 3. Maximum vertical deck accelerations — pitch and bounce test

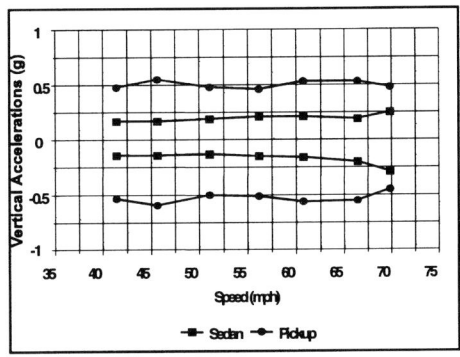

Exhibit 4. Maximum automobile vertical acceleration — pitch and bounce test

5.2.4 Twist and roll

Twist and roll response was initiated by testing on tangent track constructed with ten 39-foot vertical bumps, each 0.75 inch amplitude, out of phase on each rail. Speeds tested were from 10 to 70 mph. The criterion for success is that maximum vertical and lateral deck acceleration must be no more than ± 0.5 g. While the premium and advanced bogie designs met the requirements, the old technology bogie did not. Exhibit 5 shows the automobile response to lateral deck acceleration during the premium bogie tests.

The pickup bogie had roughly twice the response as the sedan in lateral acceleration, reaching peak response 5 to 10 mph sooner.

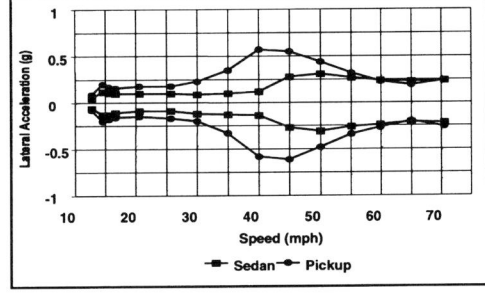

Exhibit 5. Maximum automobile lateral acceleration during twist and roll tests

5.3 Summary Of Controlled Test Results

Differences measured during controlled tests suggest that ride quality performance can be improved with better suspensions. The important issue remains: will these controlled test results predict improvements in railroad service?

5.4 Railroad Service Ride Quality Testing

In addition to controlled test requirements, the recommended practice describes a 5,500-mile railroad service test with vertical, lateral, and longitudinal acceleration measured at each deck. The test route is as follows:

- Newark, New Jersey to Chicago, Illinois on Conrail
- Chicago to Milpitas, California on Union Pacific
- Los Angeles, California to Chicago on Burlington Northern Santa Fe

Criteria for success are the number of occurrences at a predetermined level for each acceleration. In the vertical plane, one occurrence at 1.0 g or 100 occurrences at 0.5 g per thousand miles would exceed the criteria. In the lateral plane, one occurrence at 0.75 g or 100 occurrences at 0.35 g per thousand miles would exceed the criteria.

Testing in railroad service can be dominated by local factors such as train handling, performance of adjacent cars, weather, special track work, and train speed. To minimize trip-to-trip variations, the three suspensions were tested in three identical autoracks coupled together. The railroad service test program demonstrated the following:

- Performance differences between the three bogie types measured in controlled tests were confirmed by differences measured in railroad service.
- The advanced bogie met requirements for controlled testing and railroad service testing.
- The premium bogie performed better than the old technology bogie but did not meet vertical performance standards in controlled tests or in railroad service tests.
- The old technology bogie did not meet performance requirements in either test
- Improved performance in controlled testing was reflected in reduction in the number of railroad service acceleration events as well as a reduction in the maximum acceleration.

5.4.1 Vertical Performance in Railroad Service

Table 1 lists the vertical performance for the 5,578-mile trip. The primary difference between trip segments is higher train speed for the second segment. The advanced bogie met the requirements for all three trip segments. The old technology bogie did not meet requirements for any of the three segments. Although the premium bogie performed better than the old technology bogie, it also did not meet requirements for any of the three segments. These results confirm observations made during the controlled test program.

A histogram of vertical performance of the three bogie types is shown in Exhibit 6 (upper deck). The number of vertical events reduced when progressing from old technology to premium to advanced suspension types. Importantly, this trend was true at all levels of acceleration, not just at the highest *g* levels. Automobiles shipped with advanced bogie technology will be subjected to fewer acceleration events and will not be subject to high amplitude events.

Table 1. Vertical ride quality performance

* Events / 1,000 Miles	Test Bogies		
	Baseline	Premium	Advanced
Newark – Chicago			
Deck 1 > 0.5*	111	46	4
Deck 1 > 1.0	0	0	0
Deck 2 > 0.5*	282	166	19
Deck 2 > 1.0	2	0	0
Chicago – Milpitas			
Deck 1 > 0.5*	1,269	191	23
Deck 1 > 1.0	59	2	0
Deck 2 > 0.5*	2,755	452	50
Deck 2 > 1.0	170	6	0
Los Angeles–Chicago			
Deck 1 > 0.5*	23	14	0
Deck 1 > 1.0	0	0	0
Deck 2 > 0.5*	985	54	13
Deck 2 > 1.0	24	1	0

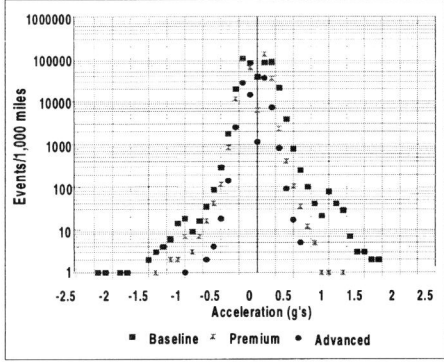

Exhibit 6. Vertical acceleration events per 1,000 miles — upper deck events>1.0 g not normalized

5.5 Summary Of Railroad Service Test Results

This railroad service test demonstrates that ride quality expectations can be met using new bogie designs. The results lend credibility to the controlled test program's ability to predict good performance in railroad service. These results also indicate that improvement to the vertical performance of the Swing Motion bogie is needed.

6. IMPROVING THE PREMIUM BOGIE

Before automobile manufacturers defined new requirements for ride quality, the NACO Swing Motion bogie was applied to several thousand bi-level autoracks to improve high-speed stability performance. However, this application does not meet new manufacturers goals for vertical ride quality performance. Can vertical performance be improved by introducing minor changes to the premium suspension? Improving the Swing Motion vertical performance was possible when TTX lowered the maximum spring capacity and maximum weight on rail for bi-level autoracks. This action was taken following a study of maximum automobile weights on bi-level railcars. The new guidelines enabled NACO to change the suspension to provide a softer vertical ride.

A standard Swing Motion bogie (designated "Bravo") and a modified Swing Motion bogie (designated "Charlie") were tested at TTC. As a result of the improvements measured, the same bogies were tested together in railroad service.

The premium bogie test program demonstrated the following:

- Improvements measured in controlled tests were proven in railroad service.
- The Swing Motion Charlie bogie reduced vertical acceleration events by 50 percent in railroad service as compared to Swing Motion Bravo.
- Vertical acceleration performance was reduced at all speeds for Charlie.
- Bravo did not meet vertical test criterion at speeds above 55 mph
- Charlie did not meet the criterion at speeds above 65 mph.
- Both Swing Motion bogie types met lateral performance criteria.
- Increasing Charlie friction damping resulted in unacceptable high-speed stability.

Although the Swing Motion Charlie does not fully meet the vertical ride quality criteria, improvements to existing equipment can change fleet performance sooner than improvements made by phasing in new suspensions.

6.1 Bogies Tested

Exhibit 7 depicts the Swing Motion bogie used in ride quality tests.

The NACO Swing Motion Bravo bogie: This Swing Motion bogie was described earlier. The secondary vertical suspension utilizes variable friction damping provided by wedges controlled by two No. 49427 coils. Low friction material is applied to the vertical surface of the wedges. The spring nest consists of six D7 outer coils.

The NACO Swing Motion Charlie bogie: The Charlie version of this bogie has four D7 outer coils per side. A further test series was conducted using iron friction wedges (designated Iron Charlie).

6.2 Controlled Tests at TTC Per RP803-96
6.2.1 High-speed stability
The Bravo and Charlie versions met the criterion for this test. The Charlie bogie was further modified with iron friction shoes (designated Iron Charlie) in an attempt to improve vertical

Exhibit 7. NACO Swing Motion bogie

performance in the pitch and bounce tests. The change to iron shoes resulted in unacceptable high-speed stability performance. The low friction surface on the friction shoes provided better ride quality performance by keeping the suspension from locking up.

6.2.2 Pitch and bounce
Exhibit 8 shows the maximum and minimum rack acceleration for the Swing Motion Bravo and Charlie bogies at each speed tested in the pitch and bounce zone.

The Charlie bogie shows significant improvement at all speeds. At speeds above 55 mph, the Bravo bogie did not meet the RP criterion. At speeds above 65 mph the Charlie bogie did not meet the criterion.

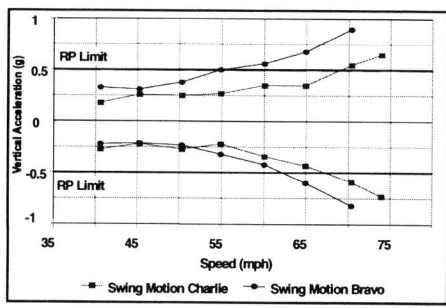

Exhibit 8. Vertical acceleration response in pitch and bounce Swing Motion — Charlie, Bravo

6.2.3 Twist and Roll
Both Swing Motion versions met criteria.

6.3 Summary of Controlled Testing
The modified Swing Motion did not meet the controlled test criteria for RP803-96. Vertical acceleration at 70 mph was reduced from 0.9 g to 0.6 g without sacrificing the lateral performance. Will this modification result in improved performance in railroad service?

6.4 Railroad Service Ride Quality Testing
To determine if the improvements noted in the controlled tests result in improvements in railroad service, an railroad service test was performed with the Bravo and Charlie cars coupled together.

6.4.1 Vertical performance in railroad service

Table 2 lists the vertical performance for the three trip segments. Due to a data loss in the final segment for the Bravo equipment, data from the 1996 Bravo test is included.

The Charlie version met criteria for two of the three trip segments. In the middle segment, the Charlie version did not meet the events above 0.5 g criterion but did meet the maximum acceleration criterion. As before, the Bravo version did not meet criteria for any segment. Two of the Bravo test segments met the events above 0.5 g criterion; but did meet the maximum acceleration criterion. The middle segment has higher speeds.

Table 2. Vertical Ride Quality Performance Compared to Criteria,
Swing Motion Bravo and Charlie Bogie Types

* Events / 1,000 miles	Swing Motion Bi-level		
	Charlie–97	Bravo–97	Bravo–96
Newark – Chicago			
Deck 1 > 0.5*	5	18	22
Deck 1 > 1.0	0	0	0
Deck 2 > 0.5*	6	26	85
Deck 2 > 1.0	0	1	0
Chicago – Milpitas			
Deck 1 > 0.5*	91	202	104
Deck 1 > 1.0	0	2	2
Deck 2 > 0.5*	126	229	259
Deck 2 > 1.0	0	3	5
Los Angeles – Chicago			
Deck 1 > 0.5*	5	N/A	4
Deck 1 > 1.0	0	N/A	0
Deck 2 > 0.5*	25	N/A	25
Deck 2 > 1.0	0	N/A	2

7. RESULTS 1996-1997

In total, 15 concepts were tested in response to solicitation for new suspensions. In addition to the Krupp TI-7R suspension, three other designs met criteria in testing at TTC. These are the GG&W bogie, the Buckeye GSI-BX bogie, and the NACO Axle Motion II bogie. A second version of the Krupp TI-7R with friction damping replacing the hydraulics also met criteria.

7.1 Advanced Bogies Tested

Advanced bogies tested in 1996-97 used Miner TCC II-60 long travel constant contact side bearings except the GG&W design. The bogie structure was designed for a 220,000-pound capacity (two bogies) but are sprung for a maximum weight on rail of 153,000 pounds.

The GG&W Bogie: The GG&W bogie (Exhibit 9) is a passive steering design based on the patented Scheffel frame mounted shear stiffener concept. The bogie is a bolsterless design using a 'Watts' connection to the center pin. Primary suspension is provided through Metacone elastomer units between the axle and frame. Secondary suspension is mounted in two load-bearing struts each consisting of one 23.5-inch coil spring and a hydraulic damper located between the frame and car body. Each strut

Exhibit 9. The GG&W Bogie

replaces a side bearing. No load is carried at the center pivot location. Longitudinal forces are transmitted to the car body through the Watts linkage. Vertical load equalization is provided by flexibility in the frame and the secondary suspension struts. Lateral stability and axle yaw motions are controlled by a linkage called a frame mounted shear stiffener.

The Buckeye Steel Castings GSI-BX Bogie: This is a modified baggage car bogie with an H-frame structure. The secondary suspension on each side sits on an equalizer beam that transfers loads between axles and to the H-frame. Each secondary suspension has two coil springs and a rotary hydraulic damper (not shown) connected between the H-frame and the equalizer beam. The coil spring was B33010-98. The rotary damper gives twice the vertical damping on the upward stroke as compared to the downward stroke. The bolster is suspended from the H-frame by swing hangers providing lateral suspension. (Exhibit 10).

Exhibit 10. The GSI-BX Bogie

The NACO Axle Motion II Bogie: The axle motion bogie is adapted from the NACO uni-truck single-axle suspension design. Two single-axle suspensions are attached in a fabricated H-frame. The primary suspension at each wheel is four 51939-1 coil springs. Two of the four springs also provide the column load for 60-degree wedges located on each side of the wheel. The primary suspension and swing hanger connection allows each axle limited longitudinal, lateral, and roll movement. This feature provides steering capability and lateral de-coupling for high-speed stability (Exhibit 11).

Exhibit 11. NACO Axle Motion bogie

8. CONCLUSION
Work performed in the Advanced Truck program has demonstrated that ride quality improvements can be made to existing automobile carrying rail cars by using advanced suspension designs. Advanced suspensions meet ride quality requirements as defined by automobile manufacturers. Tuning existing (premium) designs for the requirements of autorack service results in improved performance.

9. ACKNOWLEDGMENT
The authors acknowledge the TTX Company and members of the automotive and railroad industry for their support in promoting the concept of advanced suspensions for rail cars used in automobile transport.

The Institution of Mechanical Engineers is a leading forum for the exchange of knowledge and expertise in the field of mechanical engineering.

A wide range of events is organized by the IMechE, to which all are welcome to attend. For further information about IMechE Conferences, Seminars, Workshops, and other events please contact us for further information.

Visit our website   www.imeche.org.uk

Telephone   +44 (0) 171 222 7899

Fax   +44 (0) 171 222 4557

Or write to   Conferences and Events
                    Institution of Mechanical Engineers
                    1, Birdcage Walk
                    London
                    SW1H 9JJ

We look forward to seeing you at our future events.